CELL AND MOLECULAR BIOLOGY

£1·90·

CELL AND MOLECULAR BIOLOGY

AN APPRECIATION

EUGENE ROSENBERG

University of California
Los Angeles

HOLT, RINEHART AND WINSTON, Inc.
New York   Chicago   San Francisco   Atlanta
Dallas   Montreal   Toronto   Sydney

Copyright © 1971 by Holt, Rinehart and Winston, Inc.

All rights reserved

Library of Congress Catalog Card Number: 70–136775

SBN: 03–085312–5—Paper

SBN: 03–086296–5—Cloth

4 3 2 1   090   9 8 7 6 5 4 3 2 1

Printed in the United States of America

Every daughter gets to be like her
mother; that is her triumph.

To Leah, Robin, Stephanie, and Denise

PREFACE

This book is intended for use as part of an introductory course in biology, primarily for students who are majoring in the arts, humanities, or social sciences. Because we live in what has been called the "golden age" of science, most American universities have decided that all of their students should obtain at least a basic knowledge of science. The practical outcome of this decision has been the establishment of a number of introductory courses in the natural sciences for nonmajors. Without a clear definition of goals, often these courses are simply "watered-down" versions of the existing courses designed for the major, the amount of simplification depending on the background and capacity of the student. It has been my experience in teaching biology to nonmajors that such courses are largely irrelevant to students who do not take subsequent courses. What the nonmajor wants and needs, in my opinion, is to have his interest in biology aroused and then to be provided with a framework on which he can continue his education in biology when he leaves the university.

More specifically, I have sought to share with the student my own feeling of excitement and intellectual joy in the great discoveries of science. The approach is to select topics, such as the origins of life and molecular genetics, which are particularly interesting to almost all students, and then follow the subject from its earliest conception to its latest development, always emphasizing the observations and reasoning behind each idea and the critical experiments that were performed to test them. In this way the student can obtain what he needs most—an attitude of scientific inquiry. For it is precisely in the substitution of evidence for dogma, as a basis for belief, that science has made its greatest offering to man. Thus the student is asked to understand the evidence, analyze the hypothesis, and be willing to change his mind as new facts are presented. In turn, the attitude of scientific inquiry and historical perspective will provide the necessary backbone for a greater appreciation of the increasing number of scientific articles which are making their way into nontechnical journals.

Whereas considerable emphasis has been placed on the general nature of scientific progress, the actual subject matter of this book is limited to cell and molecular biology. This is an area of research that has yielded dramatic progress in the last few years and that promises to continue to do so for some time to come. But cell and molecular biology is not the whole of biology. Biology is a broad subject. It includes diverse disciplines such as ecology, morphology, taxonomy, embryology, neurobiology, and physiology. We have found at the University of California that the student is better equipped to appreciate the complexity and beauty of multicellular organisms after obtaining a background in cell and molecular biology. It is my hope that the terminology and concepts derived in this book will serve as the foundation for a more meaningful study of other areas of biology.

Los Angeles   E.R.
October 1970

ACKNOWLEDGMENTS

Textbooks of this type are difficult to prepare without the close cooperation and courtesy of many individuals. I am especially grateful to my colleague, R. J. Martinez, who encouraged me to write the book and then generously gave his time to provide me with advice and criticism on the entire manuscript. I should also like to thank the following individuals for reading sections of the manuscript and making valuable suggestions: S. C. Rittenberg, E. E. Sercarz, L. J. Witkin, K. Bacon, A. Miller, F. A. Eiserling, M. J. Pickett, B. Altschuler, and A. Minton.

Although specific credit is given in the legends accompanying the figures, the author expresses his sincere thanks to numerous individuals for permission to use these illustrations. Special appreciation is extended to my colleagues, K. Bacon, F. A. Eiserling, and F. S. Sjöstrand, for preparing electron micrographs specifically for this book. Appreciation is also extended to B. Altschuler, who took a number of photographs used in this book.

I acknowledge deep gratitude to my wife Leah, who read each chapter from the point of view of the lay reader. Being unfamiliar with the jargon of science, she served as both my guinea pig and my guide. On the secretarial side, I am particularly indebted to P. Tenoso and L. P. Rosenberg for typing the original manuscript and its many revisions.

CONTENTS

CELL AND MOLECULAR BIOLOGY

1
THE ORIGIN OF LIFE

The past, the finite greatness of the past! For what is the present, after all,
but a growth out of the past.

WALT WHITMAN

Genesis has an obvious and compelling fascination for all humanity. The question of how life came to be is fundamental to theology and philosophy as well as to the natural sciences, for it is only possible to understand the essence of life when we understand its origin. Recent progress in science and technology makes this question more timely than ever. Especially significant is the more detailed understanding of the essential features of existing life which has been obtained during the last 20 years. Also, as man begins his adventure into space, the possible existence of extraterrestrial life brings new significance to this age-old question.

Rather than presenting only the current views on the origin of life, several of the explanations that have been put forth in the past will be discussed. This is being done for two reasons. First, the essence of the scientific method is learning from experience. Second, it provides an opportunity to demonstrate how clear thinking and experimentation on nonpractical problems can lead to discoveries of great potential significance.

The following theories which were proposed to account for the origin of life will be analyzed in historical sequence: *spontaneous generation, continuous life, cosmozoa,* and *chemical evolution.* Emphasis will be placed on observations and experiments that were performed to test each of these theories. In this way the student can obtain what he needs most—an attitude of scientific inquiry. The substitution of evidence for dogma, as a basis for belief, is the greatest offering that science has bestowed on man. In this tradition the student is asked to accept nothing on faith. The student should understand the evidence, analyze the hypothesis, and be willing to change his ideas as new evidence is presented.

1

1•1　The theory of spontaneous generation of animals and plants

Although the theory of spontaneous generation (abiogenesis) can be traced back at least to the Ionian school (600 B.C.), it was Aristotle (384–322 B.C.) who presented to mankind the most complete arguments for and the clearest statement of this theory. In his *On the Origin of Animals*, Aristotle states not only that animals originate from other similar animals but also that *living things do arise and always have arisen from lifeless matter*. Aristotle's theory of spontaneous generation was adopted by the Romans and Neo-Platonic philosophers and through them by the early fathers of the Christian Church. With only minor modifications these philosophers' ideas on the origin of life, supported by the full force of Christian dogma, dominated the mind of mankind for more than 2000 years.

According to this theory a great variety of organisms could arise from lifeless matter. For example, worms, fireflies, and other insects arose from morning dew or from decaying slime and manure, and earthworms originated from soil, rainwater, and humus. Even higher forms of life could originate spontaneously according to Aristotle. Eels and other kinds of fish came from the wet ooze, sand, slime, and rotting seaweed; frogs and salamanders came from slime.

Rather than examining more closely the claims of spontaneous generation, Aristotle's followers concerned themselves with the production of even more remarkable recipes. Probably the most famous of these was van Helmont's (1577–1644) recipe for mice. By placing a dirty shirt into a bin containing wheat germ and allowing it to stand 21 days, live mice could be obtained. Another example was the slightly more complicated but equally "foolproof" recipe for bees. By killing a young bullock with a knock on the head, burying him in a standing position with his horns sticking out of the ground, and finally sawing off his horns one month later, out will fly a swarm of bees.

The more exact methods of observation that were developed during the seventeenth century soon led to a realization of the complex nature of the anatomy and life cycles of certain living organisms. Equipped with this better understanding of the complexity of living organisms, it became more difficult for some to accept the theory of spontaneous generation. This skepticism signaled the beginning of three centuries of heated controversy over a theory that had gone unchallenged for the previous 2000 years. What is most significant is that the controversy was to be resolved not by powerful arguments but by ingeniously designed, simple experiments.

1•2　The experiment of Redi

To the Italian physician Francisco Redi (1626–1698) goes the honor of being the first to test the theory of spontaneous generation by using carefully controlled experimental techniques. He put some meat in each of two jars. One he left open to the air (the control); the other he covered securely with gauze. At that time it was well recognized that white worms would arise from decaying meat or fish. Sure enough, in a few weeks, the meat was infested with the white worms *but* only in the control jar which was not covered. This experiment was repeated several times, using either meat or fish, with the same result. On closer examination he noted that common houseflies went down into the meat in the open jar, later the white worms appeared, and then new flies. Redi reported that he had observed the flies deposit their eggs on the gauze; however, worms developed in the meat only when the eggs got to the meat. He therefore concluded from his observations that the white worms did not arise from the putrid meat. The worms developed from the eggs which the fly deposited. The white worm then was the larva of the fly, and the meat served only as food for the developing insect.

Redi's experiment provided the impetus for testing other well-established recipes. In all cases that were examined carefully it was demonstrated that the living organism arose, not by spontaneous generation, but from a parent. Thus it was shown that the theory of spontaneous generation was based on a combination of the weakness of the human eye and bits and snatches of information gathered by accidental observation. The early biologists had seen earthworms coming out of the soil and frogs emerging from the slime of pond water, but they had not been able to see the tiny eggs from which these organisms arose. Because their observations had not been systematic, they had not seen how the mice invaded the grain bin in search of food, so they thought that the grain produced the mice. Based on the more exact methods of observation, the evidence which supported the theory of spontaneous generation of animals and plants was largely demolished by the end of the seventeenth century.

1·3 The discovery of microorganisms

Just as the theory of the abiogenesis of higher organisms was being refuted, the controversy was reopened, more heated than ever, because of the discovery of microorganisms by Antony van Leeuwenhoek. Before proceeding to the controversy over the spontaneous generation of these microorganisms, a brief account of the discovery of microorganisms is presented. Much of the material in this section is based on a fascinating biography of Leeuwenhoek by Clifford Dobell (see Suggested Readings at end of the chapter).

As a young bacteriologist, Dobell was especially interested in studying the microbial flora of the mouth. However, each time he presented his Professor with what he thought was the discovery of a new type of microbe, his Professor would shake his head and respond, "No, no, Leeuwenhoek already discovered that one." Finally, motivated by a mixture of curiosity and skepticism, he decided to find out more about this man Leeuwenhoek. After 25 years of painstaking research, in 1932 Dobell published a truly inspiring biography of Leeuwenhoek. Those students who find time to read Dobell's book will be treated to a masterpiece of English biography and rewarded with an insight into the true meaning of scientific research.

Leeuwenhoek was born in Delft, Holland, in 1632. After leaving school at 16, he moved to Amsterdam and was an apprentice haberdasher for five years. He then returned to Delft, got married, and opened up a dry goods store. Later he became chamberlain of the Council of Delft with the main task of keeping the Council chambers clean and warm. Between 1653 and 1673, he developed the curious hobby of constructing microscopes. Although he was not the first to build a microscope, his instruments were the finest of his time. Equally important, however, were his almost childlike curiosity and great skill as an objective observer of nature. Leeuwenhoek patiently

FIGURE 1.1 Antony van Leeuwenhoek (1632–1723). (Courtesy of Rijksmuseum, Amsterdam.)

improved his microscopes and developed his techniques of observation for 20 years before he reported any of his results.

Finally in 1673 Leeuwenhoek sent a letter to Henry Oldenburg, the first secretary of the Royal Society of England, describing the mouth and eye of the bee as viewed through his simple microscope. This letter was followed by several hundred more over the next 50 years, each written in "Nether-Dutch" (Leeuwenhoek knew no Latin or English), sent to England, and after translation published in the *Philosophical Transactions of the Royal Society*. Although these letters describe in great detail, for the first time, many of the parts of higher animals and plants, his greatest recognition comes from his discovery of the previously subvisible world of microorganisms.

His discovery of protozoa in fresh water is described in his sixth letter, dated from Delft September 7, 1674. In his eighteenth letter, dated October 9, 1676, he described bacteria for the first time. One of his most famous letters (the 39th, September 17, 1683), which describes bacteria in the human mouth (Figure 1.2), is an excellent illustration of his charming style and accurate observational ability.

> 'Tis my wont of a morning to rub my teeth with salt, and then swill my mouth out with water: and often, after eating, to clean my back teeth with a toothpick, as well as rubbing them hard with a cloth: therefore my teeth, back and front, remain as clean and white as falleth to the lot of few men of my years, and my gums (no matter how hard the salt be that I rub them with) never start bleeding. Yet notwithstanding, my teeth are not so cleaned thereby, but what there sticketh or groweth between some of my front ones and my grinders (whenever I inspected them with a magnifying mirror), a little white matter, which is as thick as if 'twere batter. On examining this, I judged (albeit I could discern nought a-moving in it) that there yet were living animalcules therein. I have therefore mixed it, at divers times, with clean rainwater (in which there were no animalcules), and also with spittle, that I took out of my mouth, after ridding it of airbubbles (lest the bubbles should make any motion in the spittle): and I then most always saw, with great wonder, that in the said matter there were many very little living animalcules, very prettily a-moving. The biggest sort had the shape of

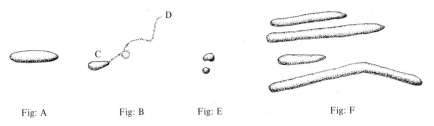

Fig: A   Fig: B   Fig: E   Fig: F

FIGURE 1.2 Leeuwenhoek's figures of bacteria from the human mouth (taken from Letter 39, September 17, 1683).

Fig. A. These had a very strong and swift motion, and shot through the water (or spittle) like a pike does through the water. These were most always few in number.

The second sort had the shape of Fig. B. These oft-times spun round like a top, and every now and then took a course like that shown between C and D: and these were far more in number.

The third sort I could assign no figure: for at times they seemed to be oblong, while anon they looked perfectly round. These were so small that I could see them no bigger than Fig. E: yet therewithal they went ahead so nimbly, and hovered so together, that you might imagine them to be a big swarm of gnats or flies, flying in and out among one another, these last seemed to me e'en as if there were, in my judgment, several thousand of 'em in an amount of water or spittle (mixed with the matter that I took from betwixt my front teeth, or my grinders).

Furthermore, the most part of this matter consisted of a huge number of little streaks, some greatly differing from others in their length, but of one and the same thickness withal; one being bent crooked, another straight, like Fig. F, and which lay disorderly raveled together. And because I had formerly seen, in water, live animalcules that had the same figure, I did make every endeavor to see if there was any life like anything alive, in any of 'em.

The most significant feature of Leeuwenhoek's work and the reason there is no question that, in fact, he did see bacteria are that his experiments are described in sufficient detail so that they can be repeated today. He tells where he obtained his material, how he handled it, and what he observed. For example, if you take the white matter between your teeth and examine it under a modern microscope, you will observe essentially what Leeuwenhoek described in 1683. Since his clear observations and correct interpretations were responsible for the discovery of bacteria and protozoa, Leeuwenhoek is recognized justly as the Father of bacteriology and protozoology.

He continued until he was 91 years old, making better lenses and watching what he called his "little beasties." He eventually was honored by being elected into the Royal Society, and was visited by such famous personages as Czar Peter the Great of Russia and Queen Mary II of England.

1·4 Needham versus Spallanzani

As soon as the discoveries of Leeuwenhoek became known, the proponents of spontaneous generation turned their attention to these microscopic organisms and suggested that surely they must have formed by spontaneous generation. Finally, experimental "proof" for this notion was published in 1749 by an Irish priest, John Tuberville Needham (1713–1781). Needham reported that he had taken mutton gravy fresh from the fire, transferred it to a flask, heated it to boiling, stoppered it tightly with a cork, and then set it aside. Despite boiling, the liquid became turbid in a few days. When examined under a microscope, it was teeming with microorganisms of all types. The experiments were repeated by and gained the support of the famous French naturalist George Louis de Buffon (1707–1788). Needham's demonstration of spontaneous generation was generally accepted as a great scientific achievement, and he was immediately elected into the Royal Society of England and the Academy of Sciences of Paris.

Meanwhile in Italy Lazzaro Spallanzani (1729–1799) performed a series of brilliantly designed experiments of his own which refuted Needham's conclusions. Spallanzani found that if he boiled the food for *one hour* and hermetically sealed the flasks (by fusing the glass so that no gas could enter or escape), then no microorganisms would appear in the flasks. If, however, he boiled the food for only a few minutes, or if he closed the flask with a cork, then he obtained the same results that Needham had reported. Thus he wrote that Needham's conclusions were invalid because (1) he had not heated his gravy hot enough or long enough to kill the microorganisms and (2) he had not closed the flask sufficiently to prevent other microbes from entering.

Count Buffon and Father Needham immediately responded that, of course, Spallanzani did not generate microorganisms in his flasks because his extreme heating procedures destroyed the *vegetative force* in the food and the *elasticity* of the air. Regarding Spallanzani's experiments, Needham wrote, "From the way he has treated and tortured his vegetable infusions, it is obvious that he has not only much weakened, and maybe even destroyed, the vegetative force of the infused substances, but also that he has completely degraded . . . the small amount of air which was left in his vials. It is not surprising, thus, that his infusions did not show any sign of life."

Rather than engage in theoretical arguments over the possible existence of these mystical forces, Spallanzani returned to the laboratory and performed another set of ingenious experiments. This time he heated his sealed flasks to boiling not for 1 hour but for 3 hours, and even longer. If Needham was right, then this treatment should certainly have destroyed the vegetative force. As Spallanzani had previously observed, nothing grew in these heated sealed flasks. However, when the seal was broken and replaced with a cork, the broth soon became turbid with microbes. Since even three hours of boiling did not destroy anything in the food necessary for the production of microbes, Needham could no longer argue that he had killed the vegetative force by the heat treatment.

Spallanzani continued to perform experiments, all of which led him to the conclusion that properly heated and hermetically sealed flasks containing broth would remain permanently lifeless. He was, however, unable to answer adequately the criticism that by sealing the flasks he had excluded the "vital forces" in the air that Needham claimed were also necessary ingredients for spontaneous generation. With the discovery of oxygen gas in 1774 and the realization that this gas is essential for the growth of most organisms, the possibility that spontaneous generation could occur, but only in the presence of air (oxygen), gained additional support.

The situation was brought to a crisis in 1859 when Félix Archimède Pouchet, a distinguished scientist and director of the Museum of Natural History in Rouen, France, reported his experiments on spontaneous generation. Pouchet claimed to have accomplished spontaneous generation utilizing hermetically sealed flasks and pure oxygen gas. These experiments, he argued, demonstrated that "animals and plants could be generated in a medium absolutely free from atmospheric air and in which therefore no germ of organic bodies could have been brought by air."

1·5 The experiments of Pasteur and Tyndall

The impact of Pouchet's experiments on his contemporaries was so great that the French Academy of Sciences offered the Alhumpert Prize in 1860 for exact and convincing experiments which would end this controversy once and for all. The prize eventually went to one whom many consider the greatest biological scientist of the nineteenth century, Louis Pasteur (1822–1895). Pasteur by this time was already famous for his experiments on the crystals of tartaric acid, diseases of silkworms, "diseases of wine" (sufficient reason in itself for Frenchmen to consider him a savior), and studies on fermentation. Once more the persuasive genius of Pasteur was to exert itself. Pasteur first demonstrated that air could contain numerous microorganisms.

FIGURE 1.3 Louis Pasteur (1822–1895). (Courtesy of Institut Pasteur, Paris.)

My first problem was to develop a method which would permit me to collect in all seasons the solid particles that float in the air and examine them under the microscope. It was at first necessary to eliminate if possible the objections which the proponents of spontaneous generation have raised to the age-old hypothesis of aerial dissemination of germs.

The procedure which I followed for collecting the suspended dust in air and examining it under the microscope is very simple. A volume of the air to be examined is filtered through guncotton which is soluble in a mixture of alcohol and ether. The fibers of the guncotton stop the solid particles. The cotton is then treated with the solvent until it is completely dissolved. All of the particles fall to the bottom of the liquid. After they have been washed several times, they are placed on the microscope stage where they are easily examined. . . .

From his microscopic observations Pasteur concluded that there are large numbers of organized bodies suspended in the atmosphere. Furthermore, some of these organized bodies are indistinguishable by shape, size, and structure from microorganisms found in contaminated broths. Later he showed that these organized bodies that collected on the cotton fibers not only looked like microorganisms but when placed in a sterile broth were capable of growth!

Pasteur's second series of experiments provided further circumstantial evidence that it was the microbes on floating dust particles and not the so-called vital forces that were responsible for sterilized broth becoming contaminated. In these experiments Pasteur carried sterile-sealed flasks to a wide variety of locations in France. At the various sites, he would break the seal, allowing air to enter the flask. The flask was immediately resealed and brought back to Paris for incubation. The conclusion from these numerous experiments was that where considerable dust existed, all the flasks would become turbid. For example, if he opened sterile flasks in the city, even for a brief period, they all became turbid; whereas in mountainous regions, especially at high altitudes, a large proportion of the flasks remained sterile.

His third and most conclusive experiment utilized the now famous swan-neck flask. As a result of the experiments described, Pasteur hypothesized that the source of contamination was dust. If true, then it should be possible to keep a broth sterile even in the presence of air as long as the dust is kept out. In order to test this hypothesis, Pasteur constructed several bent-neck flasks such as the swan-neck flasks shown in Figure 1.4. After introducing broth into the flask, he boiled the liquid for a few minutes, driving the air from the orifice of the flask. As the flask cooled, fresh air entered the flask. Despite the fact that the broth was in contact with the gases of the air, the fluid in the swan-neck flask always remained sterile. Pasteur reasoned correctly that the dust particles that entered the flask were adsorbed onto the walls of the neck and never penetrated into the liquid. As an experimental

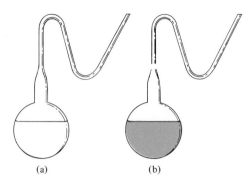

(a)   (b)

FIGURE 1.4 Pasteur's swan-neck flask. (a) Unbroken neck; nutrient medium in flask uncontaminated. (b) Flask with broken neck; contents of flask contaminated.

control Pasteur demonstrated that nothing was the matter with the broth. If he broke the neck off the flask or tipped liquid into the neck (in both cases dust would enter the broth), the fluid soon became turbid with microscopic life.

With these simple ingenious experiments Pasteur not only overcame the criticism that air was necessary for spontaneous generation but was also able to explain satisfactorily many of the sources (dust) of the contradictory findings of other investigators. Although Pasteur's conclusions gained wide support in both the scientific community and lay community, they did not convince all the proponents of spontaneous generation.

Pouchet and his followers continued to publish reports of spontaneous generation. They claimed their techniques were as rigorous as those of Pasteur. Where Pasteur failed to obtain spontaneous generation they succeeded in every case. For example, they carefully opened 100 flasks at the edge of the Maladetta Glacier in the Pyrenees at an elevation of 10,850 feet. In this region which Pasteur had found to be almost dust free, all 100 of Pouchet's flasks became turbid after a brief exposure to the air. Even when Pouchet used swan-neck flasks, he got growth. To Pasteur this disagreement no longer revolved around the interpretation of experiments; rather, Pouchet was either lying or his techniques were faulty.

Pasteur had complete faith in his own procedures and results and had no respect for those of his opponents. Thus he challenged Pouchet to a contest in which both of them would repeat their experiments in front of their esteemed colleagues of the Academy of Science. Pouchet accepted the challenge with the added statement, "If a single one of our flasks remain unaltered, we shall loyally acknowledge our defeat." A date was set, and the place was to be the laboratory in the Museum of Natural History. Pasteur arrived early with the necessary apparatus for demonstrating his techniques. Newspaper photographers and reporters were also on hand for this event of great public interest. But Pouchet did not show up, and Pasteur von by default. It is difficult to ascertain whether Pouchet was intimidated by the confidence of Pasteur or, as he later stated, he refused to partake in the "circus" atmosphere that Pasteur had created, and that their scientific findings should instead be reported in the reputable journals. At any rate, in Pouchet's absence, Pasteur repeated his experiments in front of the referees with the same results he had previously obtained. As far as the scientific community was concerned, the matter was settled. The law *Omne vivium ex vivo* (All life from life) also applied to microorganisms.

In retrospect, however, the most ironic aspect of this extraordinary contest was not that Pouchet failed to show up, but rather that if he had appeared, *he would have won!* Pouchet's experiments are reproducible. Pouchet performed his experiments in the following manner: He filled swan-neck flasks with a broth made from hay, boiled them for 1 hour, and then allowed the flasks to cool. In every flask he obtained growth. Pasteur's experiments differed only in two respects. Pasteur used a mixture of sugar and yeast extract for his broth and boiled it for just a few minutes. Pasteur never obtained growth in his swan-neck flasks. The reason for their contradictory results was not understood until 1877, 17 years later.

Mainly because of the careful work of the English physicist John Tyndall (1820–1893), Pouchet's experiments could be explained without invoking spontaneous generation. Tyndall found that foods vary considerably in the length of boiling time required to sterilize them. For example, the yeast extract and sugar broth of Pasteur could be sterilized with just a few minutes of boiling, whereas the hay infusion of Pouchet required heating for several hours to accomplish sterilization. Tyndall postulated that certain microorganisms can exist in heat-resistant forms, which are now referred to as spores. Furthermore, studies by Tyndall and the French bacteriologist Ferdinand Cohen revealed that hay infusions contain a large number of such spores. Thus the contradictory results of Pasteur and Pouchet were due to differences in their broths.

Tyndall went on to demonstrate that nutrient medium containing spores can be sterilized most easily by boiling for one-half hour on three successive days. This procedure of discontinuous heating, now called *Tyndallization*, works as follows: The first heating kills all the cells which are not spores and induces the spores to germinate (in the process of germination the spores lose their heat resistance as they begin to grow); on the second day, the spores have germinated and are thus susceptible to the heating. The third day heating "catches" any late germinating spores. Thus, with the publication of Tyndall's work, *all* the evidence which supported the theory of spontaneous generation was destroyed. Since that time there has been no serious attempt to revive this theory.

It should be pointed out, however, that by its very nature the theory of spontaneous generation cannot be disproved. One can always argue that the conditions necessary for spontaneous generation have not as yet been discovered. Pasteur was well aware of the difficulty of a negative proof, and in his concluding remarks on the controversy of spontaneous generation, he merely showed that it had never been demonstrated.

> There is no known circumstance in which it can be affirmed that microscopic beings came into the world without germs, without parents similar to themselves. Those who affirm it have been duped by illusions, by ill-conducted experiments, by errors that they either did not perceive, or did not know how to avoid.

1·6   Techniques for sterilization

Although Leeuwenhoek, working alone, was able to discover and describe numerous species of microbes, it took the combined skill and imagination of many minds to convert the science of microbiology from one solely of observation to one of controlled experimentation. What was necessary for this development was the discovery of new techniques for handling the microorganisms. In later chapters it will be shown that the ability to perform a wide variety of controlled experiments with microorganisms is a most powerful tool for probing the secrets of biology. Since many of the microbiological techniques came about as practical consequences of the controversy over spontaneous generation, we shall digress in Sections 1·6 to 1·8 from the main theme of the chapter, the origin of life, in order to describe some of the more important techniques. The most fundamental of these were techniques for sterilization.

The central issue in the controversy over the spontaneous generation of microorganisms was sterilization and subsequent contamination. Sterilization can be defined as the *complete* destruction or removal of all living organisms. Furthermore, if a substance is to remain sterile, contamination from the outside must be avoided. Primarily from the experiences gained in the last half of the nineteenth century, the following reliable techniques were developed for sterilization:

1. Steam heat: 120°C for 20 minutes;
2. Tyndallization: 100°C for 30 minutes on each of three successive days;
3. Filtration: a process of excluding microorganisms by use of filtering agents.

In the first technique, steam at 120°C is obtained by using a modified pressure cooker, called an *autoclave* (Figure 1.5). This technique is the most common, not only in research laboratories but also in hospitals and industry. Tyndallization has already been discussed in connection with the experiments of Tyndall. Filtration is widely used for sterilizing solutions that contain

FIGURE 1.5 A research autoclave. Also shown is a flask containing nutrient medium which has a cotton plug to prevent dust from entering.

heat-sensitive substances such as certain vitamins. In this process the solution is passed through a filter that has the appropriate pore size to prevent the passage of microorganisms. The filtered solution can then be transferred to a previously sterilized (autoclaved) flask. With this method direct heating of the solution can be avoided. The technique of filtration was suggested by Pasteur in 1872 when he noted that water from deep wells, which had

...dergone slow filtration through sandy soil, was essentially free of microorganisms.

Regardless of whether the microorganisms are removed by filtration or destroyed by heat, it is necessary to avoid recontamination if sterility is to be maintained. As Pasteur had so clearly demonstrated, dust particles in the air or on instruments contain microbes, and if these microbes enter a sterile solution they may initiate growth. Thus any material that comes in contact with the solution must itself be sterilized. The chief source of contamination is the air. Although swan-neck flasks prevent air contamination, they are troublesome to prepare and inconvenient for removing or adding samples. Today the most common method of preventing microbes from entering flasks or tubes which must be kept open to the air is simply to plug the opening with cotton. The cotton plug filters out the dust, and thus the microbes, but allows the oxygen in the air to enter. In this way a flask containing broth or any other solution when plugged with cotton and autoclaved for 20 minutes will remain sterile indefinitely.

As soon as these techniques for sterilization were introduced into research laboratories, they were immediately adopted for many practical operations in public health and industry. For example, the use of sterilized surgical instruments, rubber gloves, bandages, and so on, revolutionized surgery and made possible many of the dramatic advances in medicine of the past century; the entire canning industry developed as a direct result of the sterilization technique.

Many foods, such as milk, cheese, beer, and wine, cannot be sterilized without destroying their taste. In these instances a heat treatment called *pasteurization* is commonly employed. Pasteurization kills many of the microorganisms that are responsible for spoilage or for causing diseases without destroying the taste of the food. For example, (one method) the pasteurization of milk is carried out at 72°C for 15 seconds; this treatment kills the disease-producing germs in the milk but does not destroy its flavor. It should be emphasized that pasteurization is *not* a method of sterilization because many of the nondetrimental microorganisms are not killed.

1·7 Pure culture technique

The solutions that Leewenhoek and the other pioneers in bacteriology examined under their microscopes contained a wide assortment of microbes of varying sizes and shapes. Was this due to the fact that a particular organism could exist in various forms, or was it the result of a mixture of different organisms, each having a relatively fixed form? To answer this and many other microbiological questions that were being asked toward the end of the nineteenth century, it became necessary to obtain pure cultures. A *pure culture* is one that contains only a single type of microorganism. The problem then was to devise a method in which different types of microbes could be separated.

The first pure culture was obtained by the English surgeon, Lord Lister, in 1878, using a dilution technique. The principle of his method is to take a turbid culture, containing a mixed population of microbes, and to dilute it with a sterile broth until a point of dilution is reached where only *one* microbe occurs in a flask. That microbe will then multiply and give rise to a population which is derived from the single parent. Lister reasoned in the following manner: Assume that 1 milliliter (1 ml) of pond water contains 20,000 microbes of various sorts. If 1 ml of this pond water is mixed with 99 ml of sterile broth, a 1:100 dilution is obtained, which now contains approximately 200 microbes per 1 ml. When this procedure is repeated, a 1:100 of the 1:100 dilution or 1:10,000 dilution is achieved, which now contains on the average only two microbes per 1 ml. One milliliter of this 1:10,000 dilution is then divided into ten equal parts, and each part is added to sterile broth. Only two of the ten broths can become turbid because there are just two microbes in the entire 1 ml. The resulting two turbid broths would likely be pure cultures since the chance that both microbes would fall in the same 0.1 ml aliquot is only one in ten.

From this example we can appreciate some of the practical difficulties of the dilution technique. Not only does it require considerable trial and error until the correct dilution is reached, but more significantly, only the most abundant type of microorganism can be purified. Because of these reasons, the dilution technique is rarely used now as a means of obtaining pure cultures.

The method presently employed comes directly from the work of the brilliant German bacteriologist Robert Koch (1843–1910). One day Koch walked into his laboratory and noticed several colored spots on the flat surface of a slice of boiled potato which he had inadvertently left on his work bench over the weekend. Koch, in his meticulous manner, removed a bit of one of the colored spots with a sterile needle, mixed it with sterile water, and then examined it under the microscope. It was teeming with microbes. After noting that, for some unexplained reason, all the microbes from the first spot were rods of rather uniform size, he proceeded to remove and examine material from another spot. Again he noted it contained numerous microbes, but this time only spherical bacteria. Each spot he examined consisted of a different type of microorganism, but within any one spot all organisms were invariably the same. He reasoned thus: Pasteur had shown that the air contained microbes; if a single microbe from the air landed on the potato, it could use the potato for food and start multiplying; on the solid surface, the microbe could not swim far and would thus give rise to a tightly packed *colony*; since the entire colony arose from a single parent, it must be a pure culture.

FIGURE 1.6 Robert Koch (1843–1910). (Courtesy of VEB George Thieme, Leipzig.)

1•8 Streak technique

This chance observation provided the basis for the deveolpment by Koch and his co-workers of the streak technique of obtaining pure cultures. First, they prepared a series of potato slices, placed them each in a covered glass jar, and then sterilized them. These remained sterile as long as they were covered. A sterile needle was then dipped into a turbid broth culture which contained a mixture of different microorganisms. The jar was opened, and the needle containing a small droplet of the broth was streaked lightly over the surface of the potato as shown in Figure 1.7.

The organisms on the needle are thus deposited onto the solid surface. On the initial section of the streak many microbes are deposited close together so that contiguous areas of growth result. As the streaking progresses fewer and fewer organisms are deposited, until at the end of the streaking process only occasional and well-separated microbes are deposited. The process is analogous to dipping a paintbrush in a bucket of paint and then drawing the brush across a wall; at first a large quantity of paint is deposited; toward the end of the brush stroke only a few drops of paint are deposited. On incubation the microbes multiply and give rise to colonies. Those microbes near

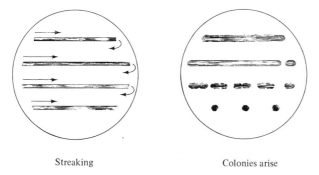

Streaking   Colonies arise

FIGURE 1.7 The streak technique for obtaining a pure culture.

the end of the streak give rise to isolated colonies that arise from a single cell and are thus pure cultures, that is, uncontaminated by other types. These can be easily transferred to sterile broth or other potato slices for further study.

Potato slices have several serious disadvantages when used for obtaining pure cultures of bacteria. The colonies tend to merge when the surface is moist; it is difficult to see colorless colonies because the potato surface is opaque; and most significant, potato is not a proper food for all bacteria. What was needed was a solid, transparent, sterile medium to which ingredients could be added as desired. Koch was able to meet these requirements by using gelatin as a solidifying agent. He simply added gelatin to the desired broth, either the yeast extract and sugar mixture of Pasteur or the hay infusion of Pouchet, autoclaved it, and then transferred the liquid into sterile dishes. When it cooled, the "jello" of bacteriologists was obtained. The gelatin medium of Koch was a major improvement over potatoes, but it also had serious defects. Gelatin is a protein, and many microbes will digest the gelatin and thus liquefy the medium. In addition, gelatin melts above 28°C and many bacteria grow best at temperatures above 30°C. Both objections to gelatin were overcome by introducing agar as the solidifying agent. Agar, an inert substance that can be extracted from seaweed, is widely used as a solidifying agent for cooking in the Orient. Once agar solidifies, it will not melt until the temperature reaches nearly 100°C, thus making it an ideal agent for bacteriological work. Nutrients can be mixed into agar to produce various "media." By utilizing the simple technique of streaking onto a solidified agar medium, the isolation of bacteria in pure culture is a routine procedure in modern laboratories.

The excursion from the main theme of this chapter, the origin of life, was undertaken for two reasons. First, the sterilization and pure culture techniques are basic to many of the discoveries which will be discussed in later chapters. Without these procedures it would have been impossible to uncover causative agents of diseases or discover methods for combatting them, such as immunization and antibiotic treatment. The use of microbes as tools for unraveling the mysteries of genetics and biochemistry depends absolutely on the development of these bacteriological procedures. Second, the development of these techniques as by-products of pure research should be emphasized. Without technology there is no science; without science there is no technology. They are mutually dependent. Through understanding we not only conquer our fear, but feed our stomachs and cure our diseases.

1•9   The theory that life is continuous

The abandonment of the theory of spontaneous generation by the end of the nineteenth century left biologists in a rather uncomfortable position. Apparently, the only alternative to spontaneous generation was the belief that life is eternal. If life does not originate from nonliving material, then it must have always been present. If this is so, the question of its origin has no meaning. Of course, it is also possible to invoke an act of supernatural creation. However, for reasons stated in Section 1•1 the concept of special creation will not be discussed. The theory that life is continuous clearly indicates that although living organisms can change their form, they can never be created from lifeless substances.

    This theory eventually came into serious question because it conflicted with certain astronomical and geochemical data. Table 1.1 summarizes the

TABLE 1.1

CHRONOLOGY OF PHYSICAL AND BIOLOGICAL EVOLUTION

EVENT	APPROXIMATE TIME (BILLIONS OF YEARS AGO)
Origin of our solar system	5.0
Formation of the earth with its present size and composition	4.5
Formation of the earth's crust	4.0
Age of oldest minerals	3.6
Earliest manifestation of life	2.7
Oldest fossils (microorganisms)	1.6
Formation of oxygen-containing atmosphere	1.0
First hard-shelled animals	0.6
Age of the dinosaurs	0.151
Earliest appearance of man	0.001

approximate time of some important events to consider in the origin and evolution of life. Most significant is the fact that the crust and the oldest minerals were formed considerably later than the earth. Both of these processes require such extreme temperatures that life in that period could not exist. At the temperatures required to melt rock in order to form the earth's crust, not only would living organisms as we now know them be unthinkable but *any* form of life based on carbon compounds would be impossible. Furthermore, the earliest manifestation of life, as evidenced by certain biogenetic calcium deposits found in Southern Rhodesia, have been dated at 1.8 billion years after the earth was formed. These and other facts strongly suggest that there was a time when the earth was sterile, and thus indicate that the theory that life on earth has been continuous is incorrect.

1•10   Panspermia theory

The evidence which contradicts the theory that life on our planet was continuous does not exclude, however, the more general theory of continuous life in the universe. Thus in 1908 the Swedish physical chemist Arrhenius revived the theory of panspermia. This theory holds that the earth is constantly being bombarded by spores from interstellar space. Once the earth had cooled sufficiently, these invading spores from other celestial bodies found the conditions favorable for growth and gave rise to living organisms on this planet.

Arrhenius made detailed calculations demonstrating how small particles could be carried upward by powerful air currents and then shot into space by electric discharges and light pressure. Once in space the spores could move with great speed. He estimated that it could take a spore "only" 9000 years to reach us from our closest neighboring solar system, Alpha Centauri. According to Arrhenius, the cold temperatures and lack of oxygen and moisture would allow the spores to survive their long trip.

Most biologists today believe that it is unlikely that spores could survive the heavy radiation of space or the frictional heat once they come in contact with the earth's atmosphere. One noted biochemist has recently suggested that, to survive such a trip, a spore would need to be fitted with a lead shield for the flight and a ceramic nose cone for the landing! The cosmozoa theory has been further criticized because it only dodges the question. If the earth was infected by spores, we would still have to explain how these spores came into existence on their native planet. For these reasons the theory of cosmozoa, although not rigorously excluded, is highly unlikely and has few advocates.

1•11   The theory of the origin of life by chemical evolution

In 1924 a Russian biochemist, Alexander I. Oparin, published a booklet which outlined his views on the origin of living matter. Later a more detailed account of Oparin's theory, documented by various investigations, appeared (1936) in his book, *The Origin of Life*. These two scholarly works revolutionized thinking on the subject and immediately provided a new experimental approach to the problem. Academician Oparin's theory of *the origin of life by chemical evolution* is described in the next few paragraphs, followed by a discussion of some recent experiments which test his theory.[1]

Oparin begins by rejecting both the theory of spontaneous generation and the theory that life is continuous. Although these theories appear to be contradictory, he clearly points out that both are based on the same dualistic outlook on nature. Both theories suggest that there is something special about life, that living organisms obey different laws from inanimate objects. The continuous life theory denotes an absolute barrier between living and nonliving; only living organisms can give rise to more life. In the theory of spontaneous generation, it was necessary to invoke a "vital force" or "vegetative force" to convert nonliving to living matter. Such a "vital force" must by its very definition be a force distinct from the usual chemical and physical forces. Oparin's theory is fundamentally different from previous theories, because it *requires no special laws for the origin of life.*

The phrase "chemical evolution" is used to emphasize the gradual, rather than the spontaneous appearance of living things. Oparin postulates a long series of chemical changes as a prerequisite to the formation of life. In this respect, his theory is much like Darwin's story of the *Origin of Species.* Both suggest a gradual transition from simple to more complex structures. Oparin's description of the chemical evolution leading to the origination of life can be considered in four phases:

Inorganic gases	$\rightarrow$	Small organic molecules	$\rightarrow$	Large organic molecules	$\rightarrow$	Aggregates or coacervates	$\rightarrow$	Primitive living organisms
	1		2		3		4	

In order to discuss these four transitions it will be necessary to introduce with each transition certain basic chemical structures. This introduction to the important chemicals of life will not only allow the reader to obtain a better understanding of Oparin's theory but will also serve as the necessary chemical background for subsequent chapters. Such a mode of introducing chemistry is a "natural" one in the sense that it attempts to follow the historical development of the chemical substances. If Oparin's theory is correct, then the chemicals will be introduced in the order of their appearance on earth.

Phase one: Production of small organic molecules

The first step in Oparin's theory is the production of simple organic compounds from the inorganic gases present in the primitive atmosphere.

[1] The English chemist J. B. S. Haldane proposed a theory very similar to that of Oparin's, which he arrived at independently, but a few years later.

Although we now know of 103 different elements, we need to consider here only six elements for our discussion of the origin and basic processes of living organisms: carbon (C), hydrogen (H), oxygen (O), nitrogen (N), sulfur (S), and phosphorus (P). These six elements comprise over 99 percent of the weight of living matter. The smallest unit in which an element can exist is called an *atom*. However, in nature matter usually exists in the form of *molecules*, combinations of atoms held together by chemical bonds. Recently the Nobel Prize winning chemist, Harold Urey, calculated that two billion years ago, when life may have arisen, the atoms of these six elements existed primarily in the following molecules:

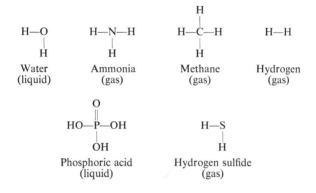

In these formulas the abbreviations for the elements shown are used, and the lines connecting the atoms represent attractive forces between atoms called *chemical bonds*. The number of bonds that a particular element forms is characteristic of that element; for example, carbon generally forms four bonds, nitrogen three bonds, oxygen two bonds, and hydrogen one bond. The chemical bonds which hold the atoms together are breakable, and if subjected to sufficient energy, can be ruptured and the atoms separated. Once some of the molecules are disrupted they can recombine to form more complex molecules. For example, if one of the hydrogens of methane is removed, it can combine with another such molecule to form the more complex organic compound, ethane gas.

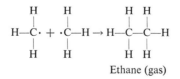

Ethane (gas)

Molecules which contain carbon are called *organic molecules* because they are characteristic of living organisms.

Urey did some further calculations which suggested that if the six molecules were present on earth a few billion years ago, then they should have formed small organic compounds as Oparin had previously speculated. Urey, at that time a professor at the University of Chicago, gave a lecture describing how various forms of energy, such as electrical storms and solar radiation, could break down these molecules and, as these bonds are rejoined, give rise to a series of more complex organic molecules. Present in his audience was Stanley Miller, a young graduate student from California. Miller was inspired by Urey's lecture and decided to put it to a test.

Over the next three years Miller planned, built, and tested the glass apparatus shown in Figure 1.8. The apparatus was designed to simulate the conditions on the primitive earth. The water was the ocean, the circulating gases were the atmosphere, and the spark was an electric storm. After first sterilizing the entire apparatus, Miller circulated and sparked the gases for several days. The electrodes were then disconnected, the gases carefully[2] removed through the stopcock, and the liquid examined for the presence of

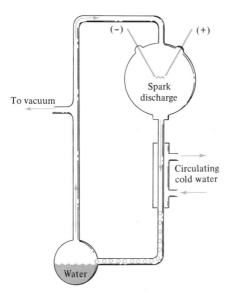

FIGURE 1.8 Apparatus designed by Stanley Miller to simulate primitive earth conditions. Methane, ammonia, hydrogen, and water vapor circulate as indicated by the arrows. After passing through the spark discharge, the gases are cooled by circulating cold water. The condensed steam containing dissolved products of the chemical reaction is collected in the flask.

[2] Any student who attempts to repeat this experiment should be warned of the potentially explosive nature of this gaseous mixture.

organic molecules. The results were exciting! A wide variety of molecules was found in rather substantial quantities. The experiment of Miller and Urey was immediately reproduced and extended in several different laboratories. The general conclusion from all these experiments was that a large diversity and quantity of organic molecules were produced when a mixture of these gases (or slightly modified mixtures) was exposed to various forms of energy. In addition to electric sparking, ultraviolet rays, visible light, heat, x rays, and ionizing radiation were successful as energy sources.

Thus the first phase in Oparin's theory now has a solid experimental basis. Small organic molecules could and *should* be formed from the mixture of primitive gases. Scientists have estimated that prior to life, the ocean was a "hot thick soup," containing from 1 to 10 percent organic molecules. These organic molecules could serve two functions: they could evolve further to more complex structures (phases 2 to 4); they could provide the first living organisms with an excellent and easily available food supply.

Of the wide variety of organic molecules identified by Miller and others in their "prebiological" syntheses, three important classes will be considered: *amino acids*, *sugars*, and *nitrogenous bases*. The chemical structure of some examples of these substances are shown in Figures 1.9 to 1.11.

Phase two: Production of large organic molecules

Oparin postulated that once these small organic molecules were concentrated in the ocean and given millions upon millions of years some of them would unite to produce larger molecules or polymers. For example, *proteins* would be formed from the union of several amino acids, *polysaccharides* from

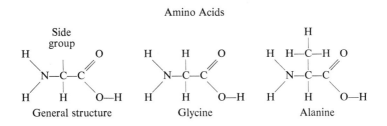

Amino Acids

FIGURE 1.9 The amino acids are composed of carbon, hydrogen, oxygen, and nitrogen atoms. Of the approximately 20 different kinds of amino acids found in proteins, two also contain S atoms. All the amino acids have the same general chemical structure, differing only in their side groups. For example, the simplest amino acids, glycine and alanine, have as their side groups an H and a CH_3 group, respectively. Amino acids are the building blocks from which proteins are constructed.

Sugars

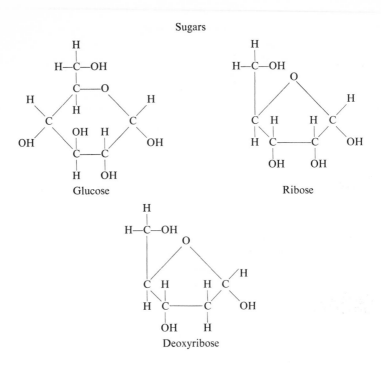

Glucose

Ribose

Deoxyribose

FIGURE 1.10 The sugars are composed of carbon, hydrogen, and oxygen atoms in a ratio close to $1:2:1$. Of the large number of different sugars found in nature, three important examples are shown. Glucose, or dextrose, which contains six carbon atoms is the most abundant sugar. It occurs in large amounts in many sweet fruits such as grapes and is also the major building block for polysaccharides. The five carbon sugars, ribose and deoxyribose, are essential components of the nucleic acids. Deoxyribose differs from ribose by having one less oxygen in each sugar molecule.

sugars, and *nucleic acids* from a mixture of sugars, nitrogenous bases, and phosphoric acid. The basic structure of these very important polymeric substances is shown in Figure 1.12.

The proteins are large molecules which contain from 50 to 1000 amino acids joined together in a long chain. The word "protein" comes from the Greek *proteios*, meaning of the first rank. These macromolecules comprise not only the major structural material of all cells but also the enzymes which promote and direct all cellular processes. These varied tasks are performed by proteins which have the same basic structure but which *differ in the number and sequence of the amino acids in the chain.* The sequence is crucial: A change in the order of the amino acids may alter the ability of the proteins to perform their varied tasks.

Nitrogenous Bases

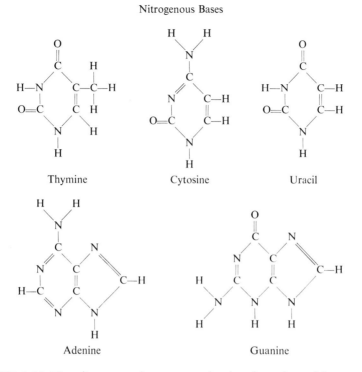

FIGURE 1.11 The nitrogenous bases are molecular rings, formed by several C and N atoms. The cytosine, uracil, and thymine rings are composed of 4 C and 2 N atoms. The larger and more complicated guanine and adenine are composed of two linked rings. The five nitrogenous bases are found in the nucleic acids, where they play the central role in the hereditary mechanism of all living organisms.

Since there are 20 different kinds of amino acids and a typical protein chain is about 200 amino acids long, the number of possible arrangements of these amino acids is astronomical, 20^{200}. It should be emphasized, however, that the actual number of different proteins found even in a simple organism such as a bacterium is about 1000. How the cell can produce the 1000 or so proteins that are required for its growth and not produce an enormous number of possible wrong proteins is a major theme of Chapter 6.

The chemical structure of the polysaccharides is simpler than that of either the proteins or nucleic acids. The three most important polysaccharides, *starch*, *glycogen*, and *cellulose*, are composed of a single kind of sugar, glucose. Plants store starch as a reserve fuel supply, whereas animals store glycogen. Cellulose is the main structural material of plants. Although it also consists of glucose units, it is useless to man as a food. Our bodies cannot break down cellulose to the usable glucose. If an economically feasible

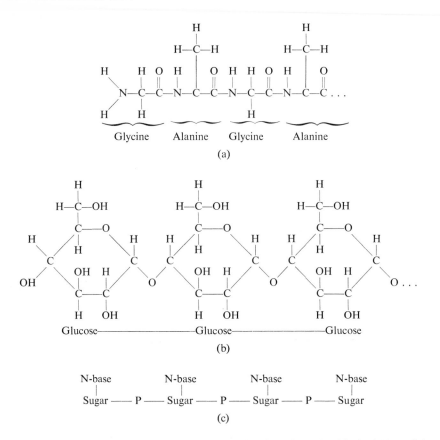

FIGURE 1.12 The basic structure of (a) protein, (b) polysaccharide, and (c) nucleic acid.

method is devised for converting cellulose to glucose, then it will be possible to change such materials as paper and cotton to sugar which could be used as food.

The nucleic acids are composed of sugars, nitrogenous bases, and phosphoric acid in equal quantities. These gigantic polymers have a main chain, or backbone, consisting of alternating sugar and phosphoric acid components; connected to each sugar residue is one of the five nitrogenous bases.

The two kinds of nucleic acids are ribonucleic acid (RNA) when the sugar component is ribose and deoxyribonucleic acid (DNA) when the sugar is deoxyribose. Both types of nucleic acids contain the nitrogenous bases, adenine, guanine, and cytosine, but the base uracil is found only in RNA, whereas the base thymine is found exclusively in DNA. Within each type of nucleic acid, the possibilities for variation are enormous. For example, although DNA molecules have the same backbone of deoxyribose and phosphoric acid, they differ in the order in which the four bases are connected to this backbone. The central role of the nucleic acids in the expression and transmission of genetic information is discussed in subsequent chapters.

A few experiments lend some support to phase two of Oparin's theory, that macromolecules can be formed from small molecules prior to the existence of life. These experiments take advantage of the fact that when the small molecules are joined together, water is produced. For example, when two glycine molecules are united, two hydrogen and one oxygen atoms (which usually combines to form water) are produced:

$$
\begin{array}{c}
\text{H} \quad \text{H} \quad \text{O} \\
\diagdown \quad | \quad \diagup\!\!\diagup \\
\text{N—C—C} \\
\diagup \quad | \quad \diagdown \\
\text{H} \quad \text{H} \quad (\text{O—H} \quad \text{H})
\end{array}
\quad + \quad
\begin{array}{c}
\text{H} \quad \text{H} \quad \text{O} \\
\diagdown \quad | \quad \diagup\!\!\diagup \\
\text{N—C—C} \\
\diagup \quad | \quad \diagdown \\
\text{H} \quad \quad \text{O—H}
\end{array}
\quad \rightleftarrows \quad
\begin{array}{c}
\text{H} \quad \text{H O H H} \quad \text{O} \\
\diagdown \quad | \quad || \quad | \quad | \quad \diagup\!\!\diagup \\
\text{N—C—C—N—C—C} \\
\diagup \quad | \quad \quad | \quad \diagdown \\
\text{H} \quad \text{H} \quad \quad \text{H} \quad \text{O—H}
\end{array}
\quad + \quad
\begin{array}{c}
\text{O} \\
\diagup\!\!\diagup \\
\text{H—O} \\
\diagdown \\
\text{H}
\end{array}
$$

Water must be also produced in the formation of polysaccharides and nucleic acids. The arrows pointing in opposite directions indicate that the reaction can also proceed in reverse. In fact, when any of these macromolecules are exposed to water, they break down to simpler organic molecules much faster than they are produced. To overcome this difficulty, Sidney Fox of Florida State University heated a mixture of *dry* amino acids at 160°C for 3 hours. The water that was produced when amino acids were thereby united was immediately boiled off and proteinlike molecules accumulated. Although the experiment was criticized by certain scientists because the temperature was too severe, Fox has pointed out that there are local "hot spots" on earth, such as hot springs and volcanoes. Furthermore, temporary dehydrating areas containing high concentration of organic molecules could be produced in tidal pools as the ocean recedes and the water evaporates.

Using a variety of dehydrating conditions, scientists have succeeded recently in producing all the different types of polymers from their respective building blocks. Although this problem is not a simple one, it is reasonable to assume that somewhere on the primitive earth during the millions of years prior to life conditions were such that the "organic soup" condensed and gave rise to polymeric substances.

Phase three: Formation of aggregates

Large organic molecules have a tendency to aggregate and form what Oparin calls *coacervates*. Coacervates can be easily prepared in the laboratory by mixing dilute solutions of various polymers. For example, on mixing a solution of gelatin (0.67 percent) with a solution of gum arabic (0.67 percent) coacervates are formed that are stable below 42°C. Coacervates appear as microscopic droplets of varying size and composition, floating in liquid medium. The formation of coacervates is dynamic; while some molecules are entering into the coacervates others are escaping back into the solution. If molecules enter the aggregate faster than they depart, then the coacervate grows in size. The increase in size makes the coacervate more susceptible to breakage by collisions or other mechanical disturbances. The fragments produced have the same composition as the original coacervate and will continue to grow and fragment (Figure 1.13).

One of the most significant features of these coacervates is that like living organisms, they occupy a definite region in space. The process of coacervate formation causes a sharp boundary to be formed for the first time between the aggregated polymers and the rest of the solution. Prior to this event the polymers were uniformly distributed in the solution. Oparin made the ingenious suggestion that in the formation and growth of coacervates a prebiological "natural selection" could operate. The various coacervates formed would compete with one another for the polymers in solution. Since the different coacervates were constantly being formed and broken down, those coacervates which were the stablest and grew the fastest would eventually predominate in the ancient seas (survival of the fittest).

Phase four: From coacervates to primitive living organisms

The coacervates have some properties which we associate with living organisms. They are able to take certain (but not all) organic matter from their environment and concentrate it within; thus they slowly become larger and eventually divide. They are, however, at the absolute mercy of the outside world in obtaining the polymers necessary for their growth. Although living organisms are also dependent on their environment for organic matter, they are able to alter the organic matter and thus provide their own building blocks. For example, when you eat a steak, the protein in the meat is broken down into amino acids which then are rebuilt into human protein. It does not matter whether the protein came from a cow, pig, fish, peanut, or any other

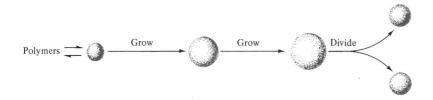

Polymers

Grow   Grow   Divide

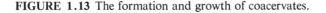

FIGURE 1.13 The formation and growth of coacervates.

source, it always ends up as distinctly human protein. The sum of all the closely coordinated, delicately balanced chemical reactions that take place in living cells, entailing the breakdown of different foods, and the synthesis of the myriad of components necessary for growth, is called *metabolism.*

> It's a very odd thing—
> As odd as can be—
> That whatever Miss T. eats
> Turns into Miss T.
> WALTER DE LA MARE

However, what happens to a coacervate when the supply of polymers in the ocean is depleted? Lacking metabolism, its growth would cease and the process of chemical evolution would come to a halt, short of the formation of living organisms. For the coacervate to cross the twilight zone between nonliving and living, it must be able to transform some of the organic material in the ocean into building blocks for its own growth. In short, it needs to develop the ability of organized self-replication through metabolism.

This is the transition step which is least understood and for which no experimental evidence exists. It must be emphasized that there is a very great gap between a coacervate which grows and divides by purely physical forces and the fantastically complex nature of a living cell. Nevertheless, Oparin argues that during the millions of years that the coacervates had to evolve, they developed through natural selection: (1) various patterns of metabolism and (2) the ability to transmit the information for these patterns from generation to generation. As these properties became established, a living organism slowly emerged.

Conclusions

The theory of the chemical origin of life has gained wide acceptance by modern biochemists and biologists. There is strong experimental evidence indicating that a variety of organic molecules could be formed on earth prior to the existence of living organisms. The transitions from these molecules to coacervates and to living organisms will require considerably more experimentation before they can be understood in detail. The most appealing feature of the theory is that it invokes no special laws. Life was a natural consequence of chemical evolution. According to Oparin, on a planet such as ours the formation of life was not only possible, but given enough time, it was inevitable.

If the formation of life is such a natural phenomenon, then it is reasonable to inquire: Is life still emerging from nonliving matter on earth? Is there life on other planets?

The formation of living organisms by chemical evolution is much less likely today than it was two billion years ago for at least two reasons: (1) the present atmosphere contains oxygen gas and (2) existing organisms would interrupt the process of chemical evolution by using the necessary organic molecules as sources of food.

As mentioned previously, there is strong evidence that the primitive atmosphere contained little or no oxygen. Oxygen gas was produced on earth as a direct result of photosynthetic organisms. Attempts to produce organic compounds from mixtures of gases containing oxygen have been unsuccessful.

According to Oparin's theory, the formation of a living organism from organic molecules is a very slow process. When the earth was sterile, these molecules could slowly accumulate. At the present time, however, the large number of microscopic organisms that populate the earth would digest such molecules and prevent their accumulation. It is an interesting historical point of Oparin's materialistic theory that once life originated, the probability of it happening again is greatly reduced. This problem was realized as long ago as 1871 when Charles Darwin wrote

> It is often said that all the conditions for the first production of a living organism are now present, which could ever have been present. But if (and oh! what a big if) we could conceive in some warm little pond, with all sorts of ammonia and phosphoric salts, light, heat, electricity, etc., present, that a protein compound was chemically formed ready to undergo still more complex changes, at the present day, such matter would be instantly devoured or absorbed, which would not have been the case before living creatures were formed.

The possible existence of extraterrestrial life is an important philosophical and scientific question intimately related to the question of the origin of life. In the next 20 years we should know whether or not life exists elsewhere within our solar system. The most promising extraterrestrial habitat is Mars. Recent experiments have indicated that certain terrestrial microbes could survive and multiply under simulated Martian conditions. Even if Mars is sterile, it will be interesting to see how far chemical evolution has proceeded. Are there small organic molecules? Polymers? Coacervates? The probability of life outside our solar system is very high. If we assume that life is a natural consequence of a given set of chemical and physical conditions, then life should have originated on any planet whose history of conditions is similar to those of earth. Present estimates indicate that there are more than 10^{10} earthlike planets in the Universe! Also, there are the additional interesting possibilities that life radically different from that on earth exists elsewhere, for example, employing silicon instead of carbon as the central element and existing in an environment of liquid ammonia rather than water.

1•12 On the nature of scientific progress

> The basic texture of research consists of dreams into which
> the threads of reasoning, measurement, and calculation are woven.
>
> ALBERT SZENT-GYÖRGI

In this chapter we have analyzed, in historical sequence, several theories on the origin of life. We have seen how the theory of spontaneous generation went unchallenged for more than 2000 years, whereas in the last 60 years our ideas on this subject have changed continuously. The mechanism which brought about these rapid changes is often called *the scientific method*. In reality, there is no such thing as *the* method of science. There is no such ideal Platonic form, which if followed, guarantees to lead you down the path of truth. The method of science is not a fixed thing. Just as our ideas on the origin of life have changed with time, so have our methods for investigating the problems. What we have as a mechanism for scientific progress is a series of operations, some manual, some mental, which in the past have proven useful for testing ideas. Many of these operations are simply refinements of the thinking of daily life. These operations can be divided into the following four categories: observations, hypotheses, experiments, and theories.

Observations

Everyone, whether he is a scientist or not, observes. The important thing is how to observe. In this respect the trained scientist differs greatly from the creative artist. The artist shapes and colors his observations with his own feelings within the bounds of his medium. He is free to abstract from his observations only those items he feels are essential and to distort their dimensions in order to emphasize certain features and arouse certain emotions. The scientist on the other hand must observe and describe natural phenomena independent of his own sentiments.

The science of bacteriology began with observations by a single man, Antony van Leeuwenhoek. His discovery of the microbial world represents an excellent example of how to observe. First, he accurately recorded the source of his material and how he handled it, in sufficient detail so that others could later repeat his observations. Next, he patiently observed the material under a variety of conditions, using the best observational aids which existed at that time. Since these aids themselves can introduce artifacts into the observations, it is essential that the observer be as familiar as possible with the tools of his trade. Remember that Leeuwenhoek studied his microscope and developed his techniques for 20 years before submitting any of his results for publication! Even when using well-established instruments, it is important to understand how they work and what their limitations are. Finally, Leeuwen-hoek recorded his observations in explicit language and as free as possible from preconceived notions.

The following short passage from Leeuwenhoek's 18th letter to the Royal Society exemplifies some of these points:

> I did now place anew about ⅓ ounce of whole pepper in water, and set it in my closet, with no other design than to soften the pepper, that I could the better study it. This pepper having lain about three weeks in the water, and on two several occasions snow-water having been added thereto, because the water had evaporated away; by chance observing this water on the 24th April, 1676, I saw therein, with great wonder, incredibly many very little animalcules, of divers sorts; and among others, some that were 3 or 4 times as long as broad, though their whole thickness was not, in my judgment, much thicker than one of the hairs wherewith the body of a louse is beset.

Notice the use of measurements by Leeuwenhoek to make his description more explicit. He does not use a pinch of pepper, but ⅓ ounce of pepper; he does not leave it for a while, but for three weeks. Since there was no lan-guage to describe the size of microbes in the seventeenth century, Leeuwen-hoek relates the size of his microbe relative to that of the diameter of a louse's hair. Leeuwenhoek realized that the hair of a louse, although it varies in length, has a relatively constant diameter.

When a science such as bacteriology evolves, it continually develops better tools that help the observer to make more accurate measurements and to extend the range of observable phenomena. Whereas Leeuwenhoek's simple microscopes had maximum magnifications of 200-fold, we now have com-pound light microscopes with magnification powers of more than 1000 and electron microscopes with useful magnifications of more than 100,000. Equally important is the development of techniques used to prepare the material for observation. For example, staining procedures have been devel-oped which aid in identifying bacteria; other staining procedures are used to make the appearance of certain parts of cells more obvious. It is important to emphasize that each discipline of science has its own characteristic tools and procedures for making observations. The further development of that discipline will depend to a large degree on how it can adapt the techniques of other disciplines to its own needs, and invent new ones.

Hypotheses

If scientists were merely satisfied with accumulation of observations, then science soon would become unwieldy and as difficult to comprehend as the Nature from which it was derived. Observations do not solve problems, but suggest them. When an observation is not satisfactorily accounted for by existing knowledge, it introduces a difficulty or a problem. The scientist then formulates a hypothesis to explain the difficulty. Precisely *how* the human mind is able to originate new thoughts (hypotheses) or combinations of thoughts is certainly not clear. Most likely, guesses, hunches, and intuition play a more important role than deductive reasoning. For this discussion, however, the important point is that a hypothesis is simply a tentative explanation to account for observed phenomena.

The discovery of microbes by Leeuwenhoek presented a new problem for biology. Where did the microbes come from? Needham and Buffon hypothesized that microbes arose spontaneously from nonliving matter. Such a hypothesis was useful (even though it was later disproven) because it focused on the problem and made certain predictions possible which could then be tested experimentally.

Experiments

Since a hypothesis is only a plausible speculation, it must be tested extensively and much evidence for it adduced before it can be accepted. The experimental process is essentially a means of testing hypotheses under controlled conditions. It is by performing experiments that the scientist attacks the problem directly. We all attempt to learn from our experiences, but the scientist attempts to experience in order to learn. In this regard it is interesting that the French word *expérience* means both experience and experiment.

An excellent example of the experimental process is Pasteur's renowned refutation of the spontaneous generation hypothesis, which predicted that sterile broth would become turbid with microbes if exposed to air. Pasteur was able to design a swan-neck flask in which the broth remained sterile even when it came in contact with air. Thus the prediction (hypothesis) was wrong. Pasteur concluded from this and other studies that the microbes arose not by spontaneous generation but from the dust in the air.

The principle of a control is fundamental in the experimental approach. The control group corresponds to the experimental group at every point except the one at issue. Pasteur's control flask was exactly like the experimental flask except that dust was allowed to enter. Immediately after sterilization the neck was broken off. Since the control became turbid with growth, it was evidence that the broth was a suitable culture medium. In general it is much easier to establish controls in microbiology than in other areas of biology. Not only can one control the environmental conditions more easily, but by using large numbers of organisms derived from the same parent (that is, pure culture), one can reduce the possibility of variations due to the living organisms themselves.

Theories

As generally used the term *theory* is applied to a hypothesis that has been extensively tested and which ties together and puts in order the results of a number of observations and experiments. Theories, however, are not the end; they are also tentative. Whenever a theory is shown to be inconsistent with an experimental result, it is the theory and not the experiment that must be discarded. Consider the following hypothetical case. Theory A exists, which explains a large number of experimental data. Another theory B is presented, which is also consistent with the experimental facts. A crucial experiment is then designed and performed in order to choose between the two theories. The results of the experiment are found to be contradictory to theory A. Thus theory A is wrong and must be discarded. This does not mean, however, that theory B is right. Some bright young scientist might think of still another theory C which is also consistent with existing facts. It then will be necessary to perform an experiment to distinguish between theories B and C.

This simple hypothetical case exemplifies the two most significant features that characterize the discipline of science. First, science is constantly changing. Since science claims no eternal truths, theories are presented with the expectation that they will need to be modified sooner or later. The method of science is logically incapable of arriving at complete and final theories. It is, so to speak, constantly under repair. Second, science has no authorities other than observations and experiments. The men of science do not ask that a theory be believed because some important authority has said that it is true. Rather, only those doctrines which are based on the facts, which can (and must) be verified by other scientists, should be accepted even temporarily. Connected with the theoretical aspects of science is technology, which can utilize the knowledge of science to produce comforts and luxuries that were impossible in the prescientific era. It is this latter aspect that gives such importance to science even for those who are not scientists.

This discussion of science and the scientific method is not meant to imply that science and its methods are any better than other fields and other methods. When the methods of science can be applied, they have provided powerful tools for understanding Nature. In the past when science came into conflict with religious creeds or authoritarian principles, it was science that was victorious. The Copernican Revolution and Evolution are two examples of these conflicts. However, the method of science is severely limited in what kinds of questions it can answer. For example, it cannot answer the question of whether or not sciences should be used for the enhancement or destruction of life. This is why Albert Einstein wrote

> Religion without science is lame;
> Science without religion is blind.

To this we add

> Science without morality is perilous.

QUESTIONS AND PROBLEMS

1.1 What do each of the following terms signify?

Tyndallization	Coacervate
Sterilization	Metabolism
Pasteurization	Polysaccharides
Pure culture	Nucleic acids
Organic chemical	Proteins
Polymer	Macromolecule

1.2 What are the similarities and differences between the theories of spontaneous generation and chemical evolution?

1.3 What were the reasons for
Pasteur using a "swan-neck flask" rather than an ordinary flask?
Miller "sparking" a mixture of gases?
Discounting the theory that life has always existed on earth?

1.4 What are the differences between elements, atoms, and molecules?

1.5 Which technique would you use to sterilize a solution of vitamins?

1.6 Give two reasons why
Needham obtained living forms, whereas Spallanzani did not.
Agar is better than gelatin as a solidifying agent.
The conditions for the slow emergence of life are no longer present.

SUGGESTED READINGS

Baker, J. J. W., and G. E. Allen II, *Matter, Energy, and Life*. Reading, Mass.: Addison-Wesley Publishing Co., Inc., 1965. An elementary text explaining the basics of organic chemistry with relevance to specific biological problems.

Bernal, J. D., *The Origin of Life*. London: Weidenfeld and Nicolson, 1967. A recent collection of works dealing with biological, chemical, and physical aspects of the origin of life.

Dobell, C., *Antony van Leeuwenhoek and his "Little Animals."* London: Staples Press, 1932. This classic biography is now available in paperback by Dover, New York, 1960.

Dubos, R., *Louis Pasteur, Free Lance of Science*. Boston: Little, Brown and Company, 1950. An interpretive biography by one of the world's foremost microbiologists and humanists.

Gabriel, M. L., and S. Fogel (eds. and trans.), *Great Experiments in Biology*. Englewood Cliffs, N.J.: Prentice-Hall, Inc., 1955. An excellent collection of classic papers, including works by Redi, Spallanzani, Pasteur, Koch, and Leeuwenhoek.

Oparin, A. I., *The Origin of Life on Earth*. New York: The Macmillan Company, 1938. A translation by Sergius Morgulis of Oparin's book first published in Moscow in 1936; the theory of chemical evolution is fully discussed. An exact reprint of the 1938 edition is available in paperback by Dover, New York, 1953.

Pauling, L., and R. Hayward, *The Architecture of Molecules*. San Francisco: W. H. Freeman and Co., 1964. An introduction to molecular architecture for the layman, including many beautiful pictures of simple and complex molecules.

Wald, G., "The Origin of Life," *Scientific American*, August 1954 (offprint 47). A popular account of Oparin's theory by a recent Nobel laureate.

2
THROUGH THE MICROSCOPE
CELLS — VIRUSES — MOLECULES*

Faith is a fine invention
When Gentlemen can see—
But *Microscopes* are prudent
In an Emergency.

EMILY DICKINSON

This chapter offers a pictorial essay of cells, viruses, and the macromolecules of which they are composed. After a brief description of the instruments which are used to visualize biological matter, photographs of microbial, animal, and plant cells are presented in order to develop one of the central themes of biology—the cell as the structural unit of all life. The component parts of the cell are then analyzed by high resolution microscopy. The subcellular structures of different organisms demonstrate some of the possible variations on the common theme of the cell.

Viruses constitute a special form of life not composed of cells. Their life cycles, exemplified by a series of high magnification electron micrographs of a bacterial virus, demonstrate their dependency on living cells for multiplication. Finally, some examples of the most important macromolecular species for both cells and viruses, the nucleic acids and proteins, are presented.

In order to best appreciate visually this collection of micrographs, representing some of the most outstanding products of the art of photomicroscopy, descriptive material has been kept to a minimum. In later chapters the nature of the structures shown here will be amplified and correlated with their biological functions.

* This chapter was written in collaboration with Dr. K. Bacon, Department of Biology, Indiana University, Bloomington, Indiana.

THE MICROSCOPE

Cytology, the study of cells, relies heavily on the use of microscopes to extend the limit of visual observation. The two major types of instruments used in research today are the light microscope and the electron microscope.

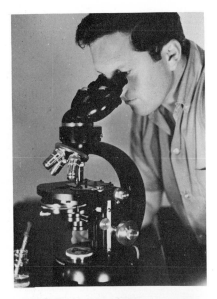

FIGURE 2.1 Light microscope
Magnifying power: 1500
Resolving power: 0.1 micron
Source of radiant energy: light
Types of specimens: living or preserved
Cost: $1000 to $2000

FIGURE 2.2 Electron microscope
Magnifying power: 200,000
Resolving power: 0.0005 micron
Source of radiant energy: electrons
Specimen preparation: preserved only
Cost: $50,000 to $100,000

RESOLUTION

The unaided human eye is capable of distinquishing two points as distinct from each other only if they are separated by at least 0.1 mm. When observing an object whose fine detail has spacings which are less than this distance, the object will appear unclear, for its fine structure will not be resolved. Merely magnifying an unclear object (for example, by photographic techniques) will result in nothing more than an unclear enlargement, blurring. The major function of the microscope is twofold: it must resolve fine detail and it must enlarge the resolved image to the dimensions that can be perceived by the human eye.

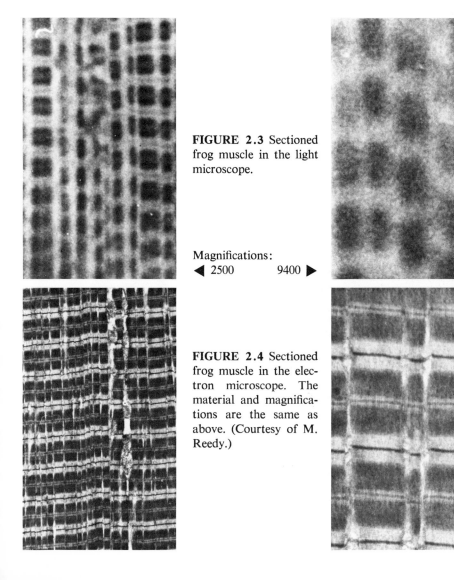

FIGURE 2.3 Sectioned frog muscle in the light microscope.

Magnifications:
◀ 2500   9400 ▶

FIGURE 2.4 Sectioned frog muscle in the electron microscope. The material and magnifications are the same as above. (Courtesy of M. Reedy.)

TWO BASIC CELL TYPES

The *eucaryotic cell* is characteristic of plants, animals, and all micro-organisms except bacteria and blue-green algae. The typical structural features include (1) a nucleus separated from the cytoplasm by a nuclear membrane; (2) cytoplasm containing many structurally differentiated units to carry out various functions; for example, respiration is localized in mitochondria and photosynthesis in chloroplasts.

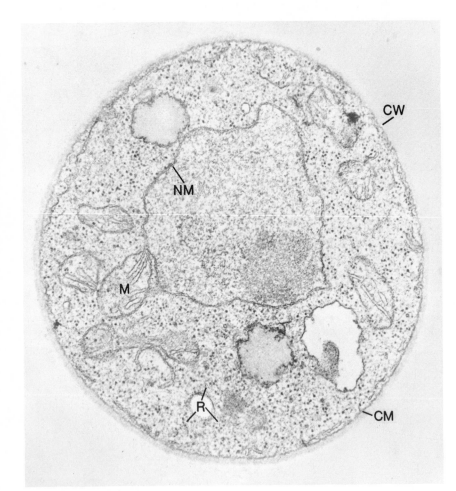

FIGURE 2.5 Thin sections of a fungus in the electron microscope. Note the nuclear membrane (NM), cell wall (CW), cytoplasmic membrane (CM), mitochondria (M), and ribosomes (R). 34,000×. (Courtesy of M. R. Edwards.)

The *procaryotic cell*, characteristic of bacteria and blue-green algae, has a relatively simple organization. The nuclear material is not separated from the cytoplasm by any defined structure and there are no distinguishable units in the cytoplasm for various cellular functions. In general, respiration and photosynthesis take place on the cytoplasmic membrane or on extensions of it.

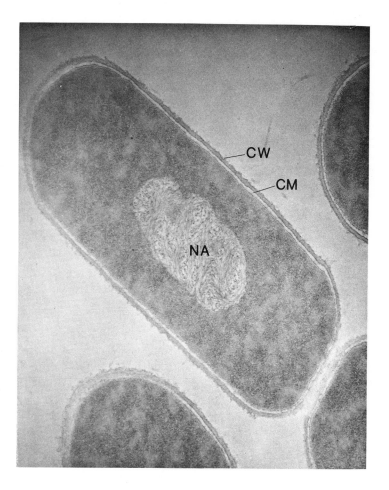

FIGURE 2.6 Thin section of the bacterium *Bacillus subtilis* in the electron microscope. Note the cell wall (CW), cytoplasmic membrane (CM), and nuclear area (NA). The darkness of the cytoplasm is the result of the presence of large numbers of ribosomes. 66,000×. (Courtesy of F. Eiserling.)

EXAMPLES OF WHOLE BACTERIA

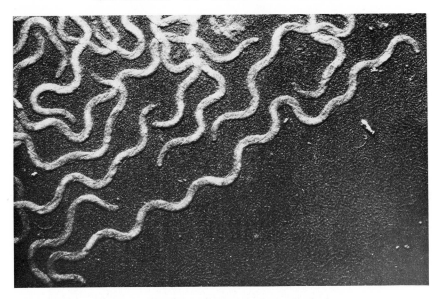

FIGURE 2.7 Spirochetes. 23,000×. (Courtesy of M. Bharier.)

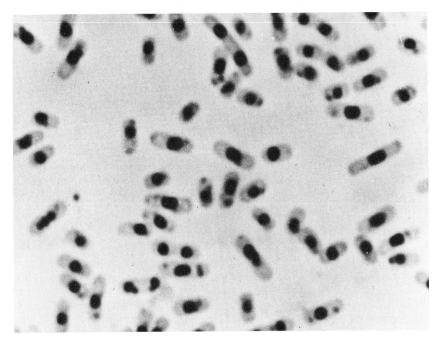

FIGURE 2.8 Rod-shaped bacteria. The nuclear regions have been stained. 1900×.

EXAMPLES OF PROCARYOTIC CELLS

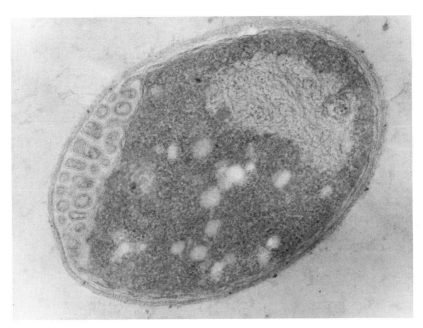

FIGURE 2.9 Electron micrograph of a thin section of a bacterial resting cell. 52,000×.

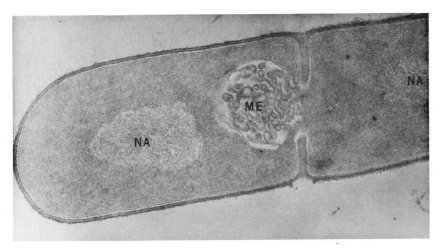

FIGURE 2.10 A thin section of *Bacillus subtilis* in the process of dividing. Note the deposition of cell wall between the two daughter cells and the presence of two nuclear areas (NA). A mesosome (ME), an invagination of cytoplasmic membrane, can be seen at the region of septum formation. 58,000×. (Courtesy of F. Eiserling.)

EXAMPLES OF EUCARYOTIC CELLS

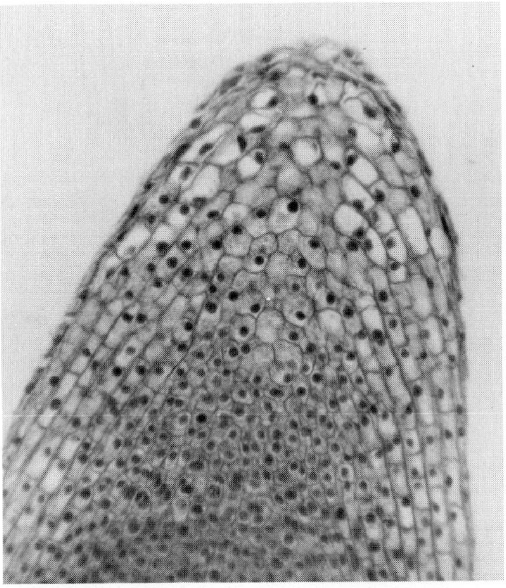

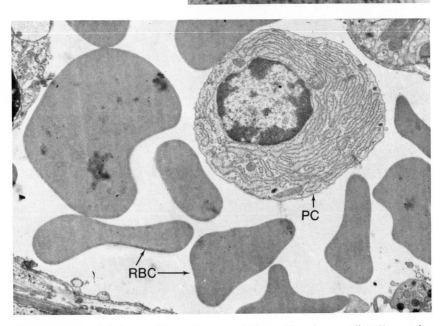

FIGURE 2.11 A section of an onion root tip in the light microscope. Note that each cell contains a darkly stained region, the nucleus. 250×.

FIGURE 2.12 Cells in a capillary of a mouse kidney. The plasms cell (PC) contains mitochondria, internal membranes, and a well-defined nucleus. The red blood cells (RBC) are exceptional in that they lack nuclei. 5800×. (Courtesy of J. Delafield.)

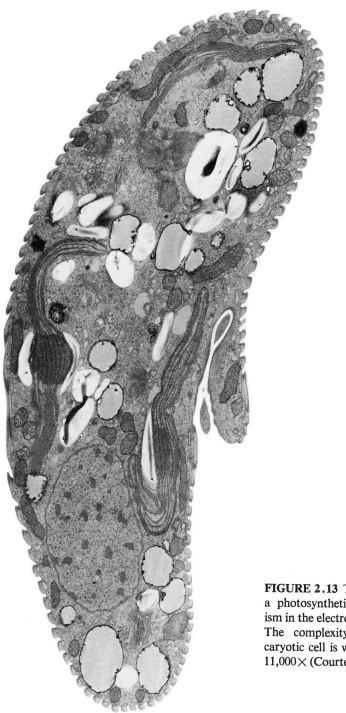

FIGURE 2.13 Thin section of a photosynthetic microorganism in the electron microscope. The complexity of the eucaryotic cell is well illustrated. 11,000× (Courtesy of J. Schiff.)

Sections of Mouse Intestine at Different Magnifications

Successive blowups of sections of mouse intestine. At low magnification, cell with cell membranes (CM), nuclei (N), mitochondria (M), and brush borders (B) are visible. Increased magnification of a selected portion of one cell reveals the finer structural details.

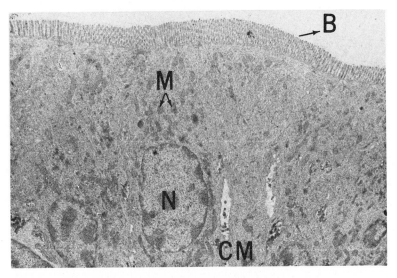

FIGURE 2.14 5100×. (Courtesy of F. Sjöstrand.)

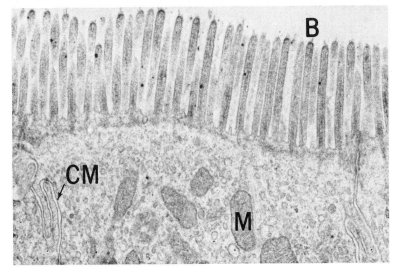

FIGURE 2.15 25,500×. (Courtesy of F. Sjöstrand.)

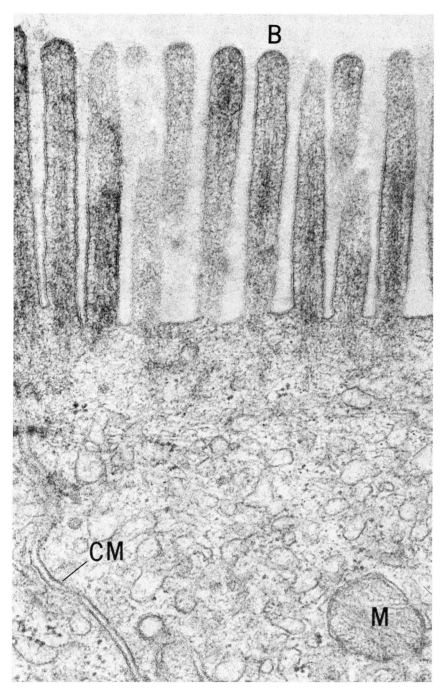

FIGURE 2.16 60,000×. (Courtesy of F. Sjöstrand.)

FLAGELLA

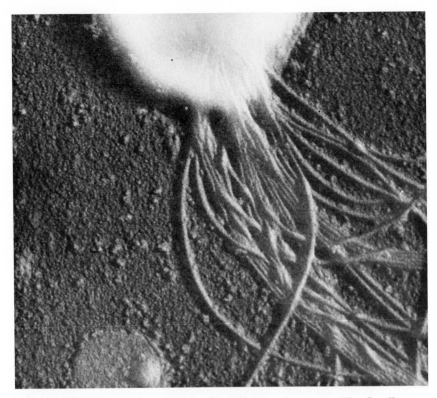

FIGURE 2.17a Bacterial flagella in the electron microscope. The flagellum, an organelle of locomotion, is composed entirely of protein. Note the flagella originating from within the cytoplasm. 214,000×. (Courtesy of R. J. Martinez.)

FIGURE 2.17b Flagella at high magnification. A cross view is indicated by the arrow. 300,000×. (Courtesy of N. Lundh.)

CELL WALLS

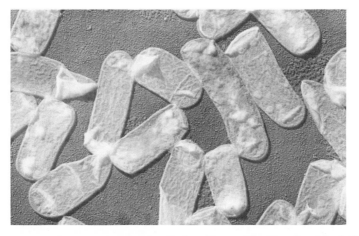

FIGURE 2.18 Electron micrograph of isolated bacterial cell walls. This rigid structure is responsible for maintaining cell shape. 60,000×. (Courtesy of E. Ribi.)

FIGURE 2.19 High magnification of a fragment of a bacterial cell wall. 70,000×. (Courtesy of R. G. E. Murray.)

MITOCHONDRION

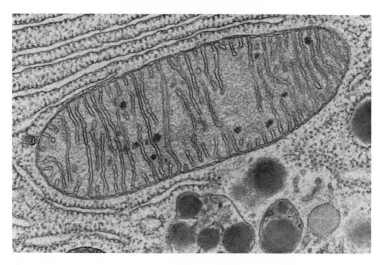

FIGURE 2.20 A typical mitochondrion from bat pancreas showing characteristic projections of membranes into the interior. As discussed in Chapter 3, the mitochondria are the major sites of energy generation in eucaryotic cells. 40,000×. (Courtesy of K. R. Porter.)

ENDOPLASMIC RETICULUM

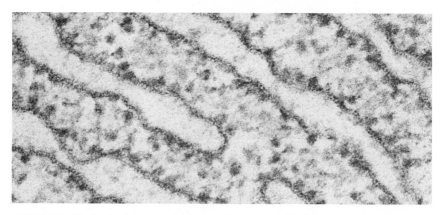

FIGURE 2.21 Endoplasmic reticulum at high magnification from sectioned mouse pancreas. This extensive system of membranes is found in the cytoplasm of eucaryotic cells. In this micrograph the membranes are seen coated with small spherical bodies called ribosomes. As discussed in Chapter 6, ribosomes are the sites of protein synthesis. Note the membranes coated with ribosomes in the upper left corner of Figure 2.20. 184,000×. (Courtesy of F. Sjöstrand.)

MEMBRANES

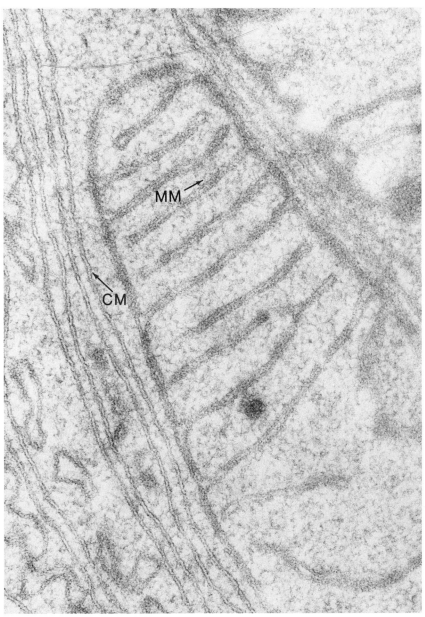

FIGURE 2.22 A thin section of mouse kidney showing two types of membrane, the layered cytoplasmic membrane (CM) and the globular mitochondrial membrane (MM). 160,000×. (Courtesy of F. Sjöstrand.)

CHLOROPLASTS: Light Microscope

FIGURE 2.23 A section of a spinach leaf in the light microscope showing four cells, each containing several chloroplasts. 15,000×. (Courtesy of I. Honda, T. Hongladrom, and S. G. Wildman.)▶

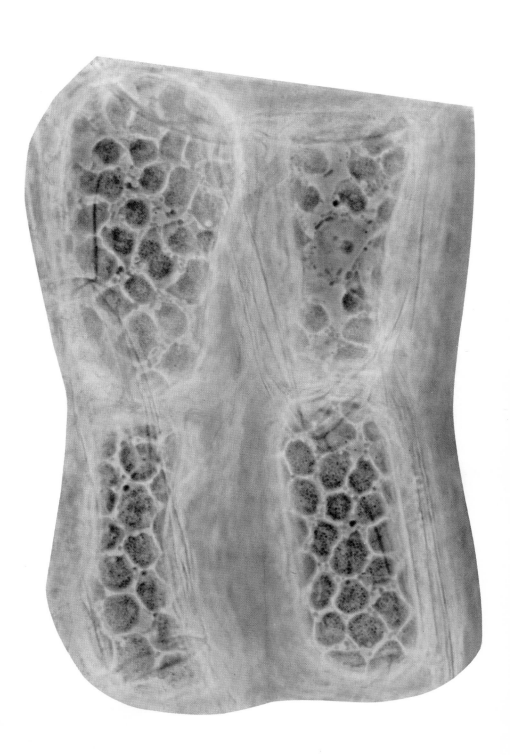

CHLOROPLAST: Electron Microscope

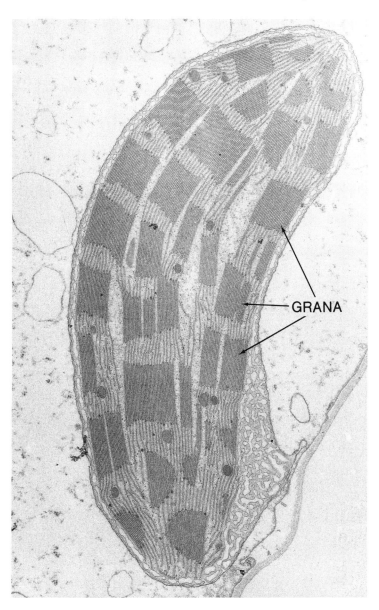

FIGURE 2.24 A section of a corn leaf chloroplast. Note the complex internal membranes which contain the chlorophyll. The heavily concentrated membranes (Grana) correspond to the dark spots visible in the chloroplast in Figure 2.23. 50,000×. (Courtesy of L. K. Shumway.)

NUCLEUS: Eucaryotic

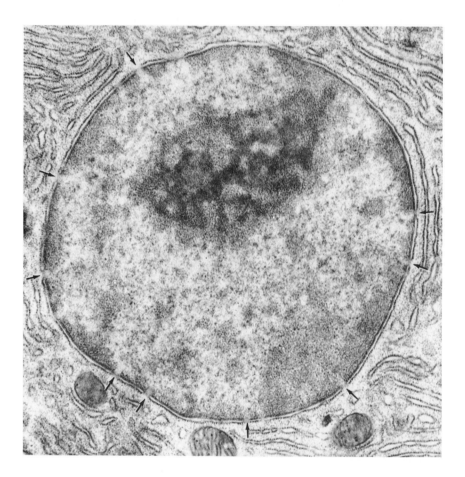

FIGURE 2.25 Electron micrograph of the nucleus from a cell of mouse pancreas. Note the pores in the nuclear membrane. The cytoplasm contains extensive membranes. 17,500×. (Courtesy of D. W. Fawcett.)

NUCLEAR REGION: Procaryotic

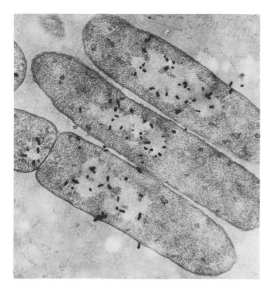

FIGURE 2.26 Thin section of several *Escherichia coli* cells in the electron microscope. Note the absence of a nuclear membrane. The dark spots on the micrograph are a result of the autoradiographic deposition of silver grains on the deoxyribonucleic acid of the cells. 22,000×. (Courtesy of F. Eiserling.)

Autoradiography. Radioactive substances, as a consequence of their emission of ionizing radiation, can, just as light, expose a photographic emulsion. This phenomenon has been exploited to make visible the location within the cell of radioactive substances used to label specific cellular structures. In the case of the localization of the nuclear material shown above, cells grown in the presence of radioactive thymine, a component of DNA, are sectioned and then covered with photographic emulsion. After an appropriate time during which the ionizing radiation in the cells exposes the emulsion, the cell sections are processed as for photography. The picture (the dark, reduced silver grains) can then be viewed in the microscope and indicates the intracellular location of the radioactive material. Thus the figure above demonstrates that DNA is concentrated in the nuclear region of the cell.

MITOSIS

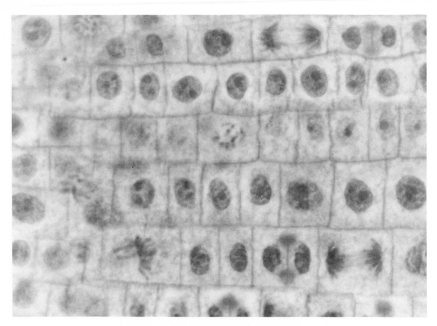

FIGURE 2.27 Onion root cells at various stages of mitosis. 1900×.

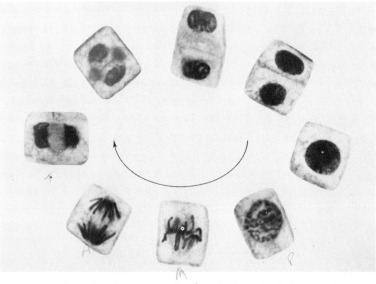

FIGURE 2.28 The ordered sequence of events during mitosis in the onion root cell. Reconstructed from micrographs like Figure 2.27.

HUMAN CHROMOSOMES

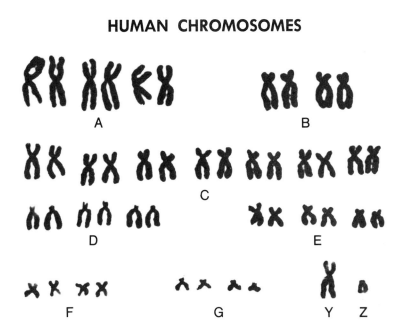

A

B

C

D

E

F   G   Y   Z

FIGURE 2.29 An array of normal male chromosomes. The chromosome pairs are arranged in groups according to size and shape. A female array would differ only in having two X chromosomes instead of one X and one Y. 2100×. (Courtesy of R. S. Sparkes.)

Abnormalities

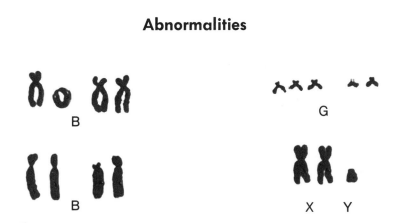

B   G

B   X   Y

FIGURE 2.30 Some chromosomal abnormalities. (Courtesy of R. S. Sparkes.) *Upper left:* ring chromosome in B group. *Lower left:* short arm chromosome in B group (cat cry syndrome). *Upper right:* extra chromosome in G group, Down's syndrome (mongolism). *Lower right:* extra sex chromosome.

VIRUSES

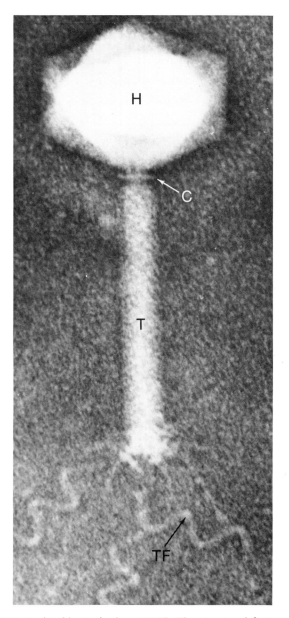

FIGURE 2.31 Portrait of bacteriophage PBSI. The structural features include the head (H), collar (C), tail (T), and tail fibers (TF). The biology of viruses will be discussed in Chapter 4. 386,000×. (Courtesy of F. Eiserling.)

LYTIC LIFE CYCLE OF A BACTERIOPHAGE

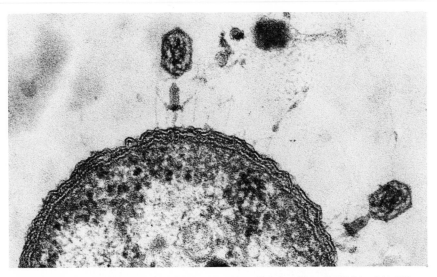

FIGURE 2.32 Adsorption of phage to cell and injection of DNA. 144,000×. (Courtesy of L. Simon.)

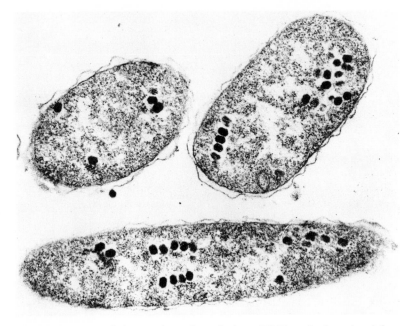

FIGURE 2.33 Intracellular condensation of phage DNA and phage head formation. 45,000×. (Courtesy of E. Kellenberger.)

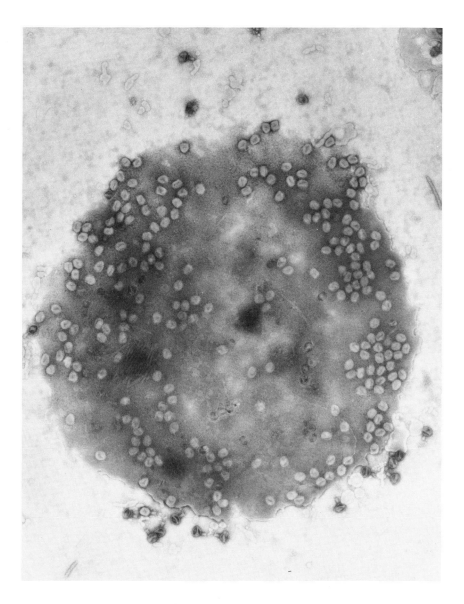

FIGURE 2.34 Lysis of a bacterial cell and liberation of the bacteriophage. The final step in the life cycle of the bacteriophage is the destruction of the bacterial cytoplasmic membrane and the release of the bacteriophage particles into the medium. 24,000×. (Courtesy of L. Cãnedo.)

SOME SITES OF PHAGE ADSORPTION

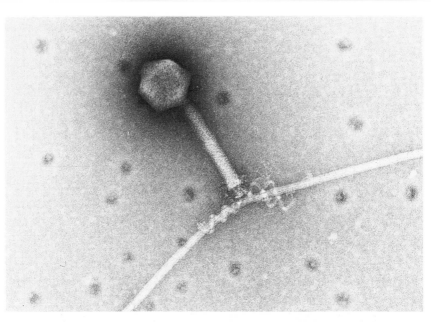

FIGURE 2.35 Bacteriophage adsorbed to a flagellum of *Bacillus subtilis*. 122,000×. (Courtesy of N. Lundh.)

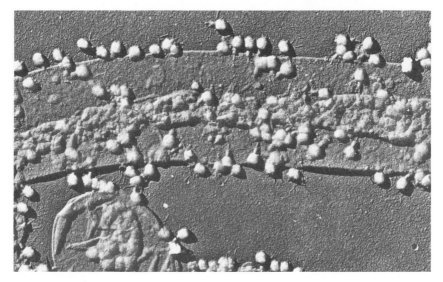

FIGURE 2.36 Bacteriophage adsorbed to isolated cell walls. 29,000×. (Courtesy of E. Kellenberger.)

PHAGE CONTRACTION

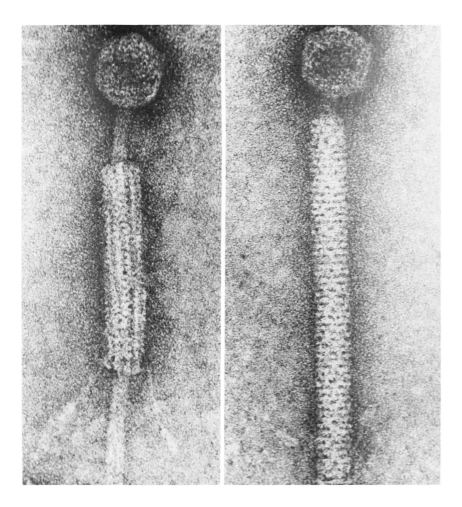

FIGURE 2.37 Electron micrograph of a *Bacillus subtilis* bacteriophage. Contraction of the tail sheath (*left*) is involved in the injection process. 547,000×. (Courtesy of F. Eiserling.)

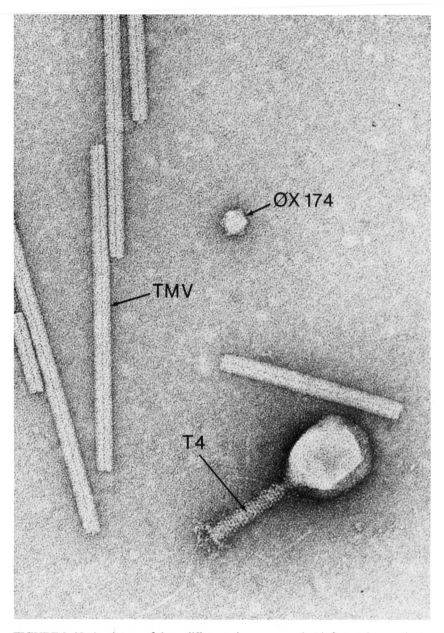

FIGURE 2.38 A mixture of three different viruses—one that infects tobacco plants, tobacco mosaic virus (TMV), and two that infect the bacterium *Escherichia coli*, bacteriophages T_4 and ϕX 174. 213,000×. (Courtesy of F. Eiserling.)

ANIMAL VIRUSES

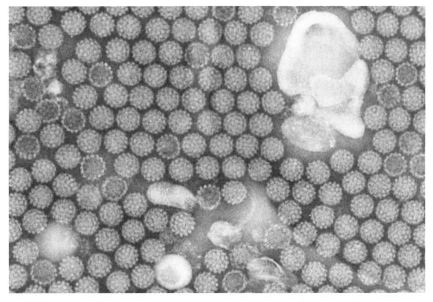

FIGURE 2.39 Human wart viruses. 116,000×. (Courtesy of A. Klug and J. T. Finch.)

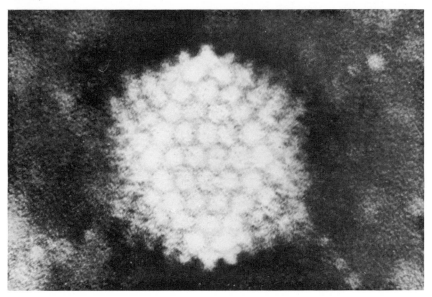

FIGURE 2.40 Adenovirus. 740,000×. (Courtesy of R. C. Valentine and H. G. Pereira.)

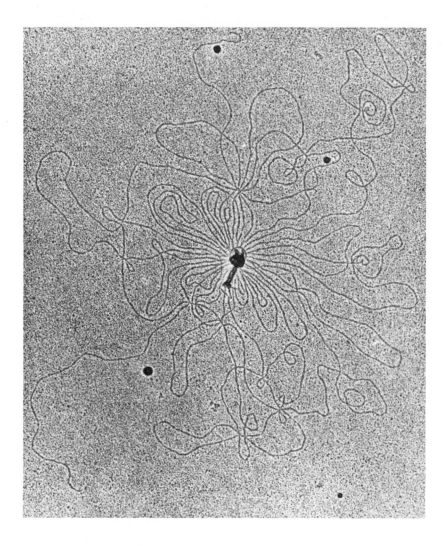

FIGURE 2.41 Threadlike DNA released from a phage head. Note the phage remnant in the center. 51,000×. (Courtesy of A. K. Kleinschmidt.)

PROTEIN MOLECULES

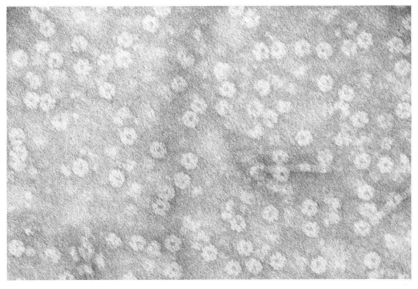

FIGURE 2.42 A doughnut-shaped molecule, glutamine synthetase. 400,000×. (Courtesy of F. Eiserling.)

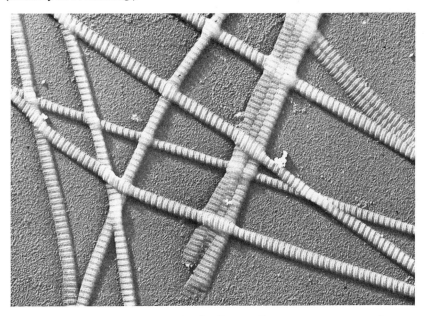

FIGURE 2.43 Electron micrograph of collagen. These structures are major components of skin and bones. 25,000×. (Courtesy of J. Gross.)

A CIRCULAR DNA MOLECULE

FIGURE 2.44 The DNA molecule from a single bacteriophage. Note that the DNA has no ends. 51,900×. (Courtesy of D. Coombs.)

3
ENERGY TRANSFORMATIONS IN BIOLOGICAL SYSTEMS

Give me matter and motion and I will make the world

RENÉ DESCARTES

As seen in Chapter 2, living material at all levels—multicellular, cellular, sub-cellular, and viral—is in a highly organized state. To maintain and reproduce these complex structures living organisms require a constant supply of energy. The series of closely coordinated, delicately balanced biochemical reactions by which the cell is able to acquire energy from various nutrients and then utilize the energy for the synthesis of cellular material is called *metabolism*. But before any discussion of metabolic reactions it is necessary to introduce first, certain thermodynamic principles and second, the concept of biological catalysis.

3•1 The first and second laws of thermodynamics

In 1847 the German physicist Hermann van Helmholtz formulated the law of *the conservation of energy*. To date there is no known exception to this law.[1] It states that the quantity of energy remains constant during the numerous changes that occur in nature. Energy can exist in a number of different and interchangeable forms, such as electrical, chemical, gravitational, light, mechanical, and thermal energy. The important point is: For any quantity of one form of energy that is lost an equivalent amount of other forms must be generated. The law of conservation of energy is also called the *first law of thermodynamics*.

First law: The energy of the universe is always constant.

[1] With the discovery by Albert Einstein in 1905 that mass and energy are interconvertable ($E = mc^2$), the law was generalized to include the conservation of both mass and energy. However, for the biological systems which will be discussed, the interconversion of mass and energy can be generally disregarded.

Considering only the first law of thermodynamics, it should be possible to build a perpetual motion machine. For example, we can imagine an isolated machine of the following type: Electrical energy is converted into mechanical energy (for example, a garbage disposal unit); the mechanical energy thus produced is utilized to generate electricity (a dynamo). Such an imaginary machine could then operate indefinitely, with no additional source of energy. In reality, however, the machine would cease to function after a few cycles. This would seem to contradict the first law since there is an apparent loss of energy in the system. The explanation, of course, is that there is less than 100 percent efficiency in the interconversion of electrical and mechanical energy. Other forms of energy such as heat are also produced. However, even if we could salvage the thermal energy and other forms of usable energy and "feed" them back into the machine, it would still cease to function after a short time. It is for this reason that physical chemists defined a new state of energy called *entropy*,[2] which is characterized by its inability to perform work. Using modern statistical physics, it is possible to demonstrate that entropy is analogous to randomness or lack of order; the greater the randomness the greater the entropy.

The *second law of thermodynamics* cannot be expressed in such concise form as the first law; it is stated in various ways according to the type of problem under investigation. To emphasize the irreversibility of real processes, the second law is stated as follows: In any real physical, chemical, or biological process, there is a net increase in entropy.

The formulation of the second law arose as a consequence of a *reductio ad absurdum*.

1. The sum of all energies in the universe is constant (first law of thermodynamics).
2. Entropy is the unique state of energy that cannot be utilized to perform work (by definition).
3. If entropy either remains constant or decreases in any series of real processes, there would be no loss in the capacity to perform work. This would lead directly to a perpetual motion machine.
4. Perpetual motion machines are contrary to experience.
5. Therefore *the entropy of the universe always increases* (second law of thermodynamics).

Since it is experimentally difficult to determine entropy directly, chemists and biologists frequently use the *free energy* to predict whether or not a reaction is energetically feasible. The free energy is that part of the total energy that can do work at a fixed temperature. Since the energy unavailable

---

[2] To be more precise, it is entropy multiplied by the absolute temperature that has the dimensions of energy.

for work (entropy) increases in any real reaction, the energy available for work (free energy) must decrease. It may be said, then, that the *direction* of a chemical reaction is determined by the sign of the free energy change; the reaction will proceed only in the direction that results in a decrease of free energy.

The second law has certain philosophical as well as scientific implications. The statement that the entropy or the randomness of the universe is always increasing points directly to a relationship between time and energy. The further we go back in history, the greater the amount of energy that was available to do work. As the philosopher Sir Arthur Eddington succinctly expressed it, the second law of thermodynamics is "time's arrow." It follows, then, that the ultimate fate of the universe is to reach a state of complete randomness or maximum entropy, which has been called *entropic doom*.

Does the second law of thermodynamics apply in biology? A fundamental characteristic of all living organisms is the ability to reproduce their highly organized nature. This would seem, at first glance, to contradict the second law—lead to an increase in orderliness, decrease in entropy. It must be emphasized, however, that the second law states explicitly that the entropy *of the universe* always increases. Thus it is necessary to think not only about the living organism but also about its surroundings. Consider, for example, a microorganism growing in an aqueous medium containing sugar and inorganic salts. The growth of the microbe is dependent on the breakdown of sugar molecules into simpler (less ordered) substances. Careful analyses have demonstrated that the total decrease in entropy of living matter during the growth process is more than offset by the increase in entropy of the nutrient molecules. Although a certain part of the system has become more highly organized, the net result is an increase in entropy. Therefore the second law of thermodynamics has not been violated.

3·2 The fermentation controversy

The laws of thermodynamics allow us to predict whether or not a particular chemical reaction is possible and in which direction it will proceed, but thermodynamics cannot tell us *when* or *how fast* the reaction will go. For example, consider the union of hydrogen and oxygen gas to produce water:

$$2H_2 + O_2 \rightarrow 2H_2O$$

Since the free energy of two water molecules is considerably less than the free energy of an oxygen and two hydrogen molecules, the reaction must proceed in the direction indicated by the arrow. However, at temperatures below 400°C, water is not formed at a measurable rate. If a small amount of finely divided platinum is added, the reaction will then proceed rapidly. Surprisingly, the platinum is not used up, but may be recovered unchanged at the end of the reaction. This is an example of a rather common phenomenon known as *catalysis*. Any substance that speeds up the rate of a chemical reaction but which itself can be recovered unchanged at the end of the overall process is known as a *catalyst*.

Much of our present knowledge concerning catalysis in living cells can be traced back to the nineteenth century controversy regarding the causative agent of *fermentation*. The phenomenon of fermentation as seen in the production of wine, beer, cheese, and bread has been known to man since antiquity. For this discussion, the production of beer will serve to illustrate the basic principles.

> Beer is made from cereal seeds, such as barley, wheat, rye, rice, and corn. After the seeds have been steeped for some time in water, they are drained and subjected to sufficient temperature to cause the moist grain to germinate; dried germinated barley seeds are called *malt*. The malt is ground, mashed in warm water, and boiled to extract the sugar and flavor from the broken cells. Hops (the dried flowers of *Humulus lupulus*) are added when the boiling is almost over, both for their bitter flavor and because the extract tends to inhibit bacterial growth. The resulting extract, the *wort*, contains a moderate concentration of sugar which has been set free from the starch in the grain. The wort is immediately cooled and large quantities of yeast are added to it; the yeast is usually obtained from a previous fermentation. Shortly after addition of the yeast, the wort starts the bubbling and frothing from which fermentation derives its name (*fervere*, to boil). After several days the reaction ceases and a copious precipitate begins to settle to the bottom of the vessel. The clarified liquid (beer) is then stored in vats at low temperatures for several weeks (lagering) prior to bottling. The sediment which has been called the ferment or yeast is used to inoculate a fresh wort.

The art of brewing was developed by trial and error over a 6000-year period and practiced without any understanding of the underlying principles. From long experience the brewer learned the conditions, not the reasons, for success. Only with the advent of experimental science in the eighteenth and nineteenth centuries did man attempt to explain the mysteries of fermentation. Let us, then, from our vantage point in time, trace the observations, experiments, and debates out of which evolved our present understanding of fermentation and biological catalysis.

For centuries fermentaion had a significance that was almost equivalent to what we would now call a chemical reaction, an error that probably arose from the vigorous bubbling seen during the process. The conviction that fermentation was strictly a chemical event gained further support during the early part of the nineteenth century when French chemists led by Lavoisier and Gay-Lussac determined that the alcoholic fermentation process could be expressed chemically by the following equation:

$$C_6H_{12}O_6 \rightarrow 2C_2H_5OH + 2CO_2$$

Glucose	Ethyl alcohol	Carbon dioxide

It was, of course, known that yeast must be added to the wort in order to ensure a reproducible and rapid fermentation. The function of the yeast, according to the chemists, was merely to catalyze the process. All chemists agreed that fermentation was in principle no different from other catalyzed chemical reactions.

Then in 1837 the French physicist Charles Cagniard-Latour and the German physiologist Theodor Schwann independently published studies which indicated that yeast was a living microorganism. Prior to their publications yeast was considered a proteinaceous chemical substance. The reason that the two workers came up with the same observations at approximately the same time is most likely due to the production of better microscopes. As Cagniard-Latour stated,

> Twenty-five years ago I first examined fresh yeast under the microscope. However, my instrument was very poor, and I concluded that the yeast was like a very fine sand composed of crystalloid particles. These observations were in error. The majority of the microscopic observations indicated in this memoir was performed on a microscope recently constructed by M. Georges Oberhauser. It enabled me to obtain enlargements of 300–400 times.

It should be mentioned that one of the reasons that it was difficult to ascertain whether or not yeast is living is because, like most other fungi, yeast is not motile. The organized cellular nature of yeast was discovered only when improved microscopes became available. Schwann and Cagniard-Latour also observed that alcoholic fermentaion always began with the first appearance of yeast, progressed only with their multiplication, and ceased as soon as their growth stopped. Both scientists concluded that alcohol is a by-product of the growth process of yeast. As Schwann stated,

> The alcoholic fermentation must be considered to be that decomposition which occurs when the sugar fungus utilizes sugar and nitrogen containing substances for its growth, in the process of which the elements of these substances which do not go into the plant are preferentially converted into alcohol. Most of the observations on the alcoholic fermentation fit quite nicely with this explanation.

The biological theory of fermentaion advanced by Cagniard-Latour and Schwann was immediately attacked by the leading chemists of the time. The eminent Swedish physical chemist Jöns Jakob Berzelius reviewed the two papers in his *Jahresbericht* for 1839 and concluded that microscopic evidence was of no value in what was obviously a purely chemical problem. According to Berzelius, nothing was living in yeast.

It was only a chemical substance which precipitated during the fermentation of beer and which had the usual shape of a noncrystalline precipitate.

To the scorn of Berzelius was soon added the sarcasm of two great German chemists, Justus von Liebig and Friedrich Wöhler. In 1839 there appeared in the *Annalen der Chemie*, a reputable scientific journal, an anonymous[3] article entitled "The Riddle of the Alcoholic Fermentation Solved." In this scientific travesty, the yeast is viewed through a powerful new microscope (Figure 3.1). When fed a solution of sugar, the yeast devours it, and a stream of alcohol is seen flowing from the anus while carbon dioxide bubbles out of its enormously enlarged genital organs. The bladder of the yeast when full has the shape of a champagne bottle.

In addition to this farce, Liebig published a serious paper, containing several important arguments against the biological theory of fermentation. Liebig's two major points can be summarized as follows:

1. Certain types of fermentation such as the lactic acid (souring of milk) and acetic acid (formation of vinegar) fermentations can occur in the complete absence of yeast.
2. Even if yeast is living, it is not necessary to conclude that the alcoholic fermentation is a biological process. The yeast is a remarkably unstable substance which, as a consequence of its own death and decomposition, catalyzes the splitting of sugar. Thus fermentation is essentially a chemical change catalyzed by breakdown products of the yeast.

Liebig's views were widely accepted, partly because of his powerful influence in the scientific world and partly because of a desire to avoid seeing an important chemical change relegated to the domain of biology. So the stage was set—biology against chemistry—for the entrance again of Pasteur.

In 1857 Pasteur published his first paper in the field of fermentation. The publication dealt with lactic acid fermentation, not alcoholic fermentation. Utilizing the finest microscopes of the time, Pasteur discovered that souring of milk was correlated with the growth of a microorganism, but one considerably smaller than the beer yeast. During the next few years Pasteur extended these studies to other fermentative processes, such as the formation of butyric acid as butter turns rancid. In each case he was able to demonstrate the involvement of a specific and characteristic microorganism; alcoholic fermentation was always accompanied by yeasts, lactic acid fermentation by nonmotile bacteria, and the butyric acid fermentation by motile rod-shaped bacteria. Thus Pasteur not only disposed of one of the opposition's strongest arguments but also provided strong circumstantial evidence for the biological theory of fermentation.

[3] It was later revealed that the article was authored by none other than Liebig and Wöhler, the editors of the *Annalen*.

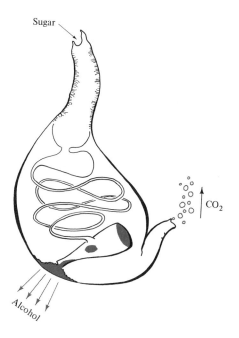

Sugar

CO_2

Alcohol

FIGURE 3.1 The yeast cell as described by Liebig and Wöhler. (Original drawing by F. W. Taylor from an English translation.)

Now Pasteur was ready to attack the crucial problem, alcoholic fermentation. Liebig had argued that this fermentation was the result of the decay of yeast; the proteinaceous material which is released during this decomposition catalyzes the splitting of sugar. Pasteur countered this argument by developing a protein-free medium for the growth of yeast. He found that yeast could grow in a medium composed of glucose, ammonium salts, and some incinerated yeast. If this medium is kept sterile, neither growth nor fermentation takes place. As soon as the medium is inoculated with even a trace of yeast, growth commences and fermentation ensues. The quantity of alcohol produced parallels the multiplication of the yeast. In this protein-free medium, Pasteur was able to show that fermentation takes place without the decomposition of yeast. In fact, the yeast synthesize protein at the expense of the sugar and ammonium salts. Thus Pasteur concluded in 1860 that

> Fermentation is a biological process, and it is the subvisible organisms which cause the changes in the fermentation process. What's more, there are different kinds of microbes for each kind of fermentation. I am of the opinion that alcoholic fermentation never occurs without simultaneous organization, development and multiplication of cells, or continued life of the cells already formed. The results expressed in this memoir seem to me to be completely opposed to the opinion of Liebig and Berzelius.

Pasteur argued effectively and, more importantly, all the data were on his side. Thus the vitalistic theory of fermentaion predominated until 1897, when an accidental discovery by Eduard Buchner finally resolved the controversy and threw open the door to modern biochemistry.

Buchner, working in Munich with his brother Hans, was attempting to obtain from yeast an extract which might have medicinal value. After several unsuccessful trials, he discovered that yeast could be disrupted and the cell sap released by grinding a mixture of intact cells and fine sand with a mortar and pestle. After filtering the mash to remove the sand and any unbroken cells, a clear yeast juice was obtained. The juice, however, soon became contaminated with bacterial growth. Since the extract was to be used for human consumption, Buchner could not utilize ordinary antiseptics to prevent the spoilage. Therefore he attempted to preserve the yeast extract by adding large quantities of sugar—much as a housewife uses high concentrations of sugar in preserving jam and jelly. To Buchner's utter amazement, the yeast extract began to bubble soon after the sugar was added. Careful analysis revealed that the sugar was decomposing to carbon dioxide and ethyl alcohol. Fermentation had proceeded *in the absence of living cells*.[4]

3·3 Biological catalysts: The enzymes

Buchner's achievement inaugurated a new era in the study of alcoholic fermentation and other metabolic processes. Reactions which normally take place only in living cells (*in vivo*) could now be studied in test tubes (*in vitro*). The agents which are present in cell extracts and which catalyze these reactions (make them go faster) are called *enzymes*. The term enzyme comes from the Greek words *en zyme*, meaning "in yeast."

During the twentieth century enzymes have been separated and obtained in pure form from a wide variety of different organisms. From studies on these purified enzymes, certain generalizations have emerged.

1. *Enzymes are true catalysts.* They control the rate of a chemical reaction, but they themselves are not used up during the process. Thus they can be used over and over again. Although enzymes can control the speed of reactions, they cannot bring about reactions that otherwise would not occur, that is, reactions involving an increase in free energy.

2. *Enzymes are large protein molecules.* Most enzyme molecules are composed of 100 to 1000 amino acids joined together in a specific sequence. In addition to the sequence of amino acids, the three-dimensional structure of the protein also plays a critical role in catalytic activity. For example, heating

[4] In 1907 Eduard Buchner received the Nobel Prize in Chemistry "for his biochemical researches and his discovery of cell-less fermentation."

an enzyme can alter the shape of the molecule, and thus its catalytic activity will be destroyed. Most likely, sterilization by heating is effective because essential enzymes are inactivated.

3. *Enzymes are highly specific in their action.* Even the simplest microorganism contains more than 1000 different enzymes, each capable of catalyzing a specific reaction. In general, a different enzyme is used to catalyze each of the different steps in the synthesis and breakdown of organic molecules that take place in living organisms.

4. *Enzymes determine the metabolic pattern of the cells.* All important reactions which take place in living cells are catalyzed by enzymes; thus species of cells must achieve their individuality from the kind and amount of enzymes they contain. For example, the ability to derive nutrient value from protein depends on the presence of a group of enzymes, called proteinases.[5] Absence of these proteinases means that the cell or organism cannot digest protein. Enzymes also determine the synthetic capacity of cells. Dark-skinned individuals have abundant quantities of the series of enzymes which catalyze production of the pigment melanin. The complete absence of any one of these enzymes results in albinism—fair skin, white hair, and pink eyes.

One of the most important areas of current biochemical research deals with the mechanism by which enzymes accelerate reactions. Although at present no theory adequately explains the phenomenon of enzyme catalysis, Figures 3.2 and 3.3 are offered as a visual representation of two possible mechanisms.

Before discussing biochemical aspects of energy metabolism, let us take one final look at the fermentation controversy in light of our current knowledge about enzymes. Who was right, Liebig or Pasteur? Pasteur was certainly wrong in his generalization that fermentation can occur only in living cells. But he was correct in his fundamental thesis that fermentation is brought about in nature as a consequence of the living activities of microorganisms. Liebig's idea about the catalysis of fermentaion by the decomposition of yeast was (and still is) inconsistent with the facts. Furthermore, his failure to recognize Pasteur's fundamental thesis was scientific blindness. In a more profound sense, however, Liebig was right in arguing that fermentation could be explained by chemistry. Today it is almost an article of faith that even the most complex biological process can be understood in chemical terms. Thus there was some value in both points of view. Fermentation is a biological process, resulting from a series of chemically intelligible reactions, each

[5] The suffix *-ase* indicates that the substance is an enzyme. The few enzymes which end with *-in*, such as trypsin and pepsin, were named before an international ruling was made favoring the *-ase* ending The root usually indicates the type of molecule on which the enzyme operates. Proteinases are enzymes which attack protein, peroxidases split peroxides, dehydrogenases remove hydrogen atoms, and so on.

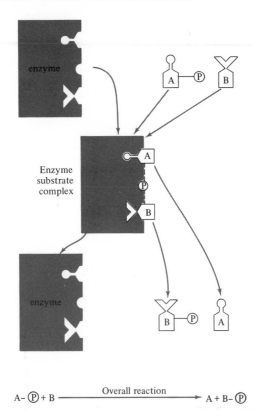

$$\text{A-}\textcircled{P} + \text{B} \xrightarrow{\text{Overall reaction}} \text{A} + \text{B-}\textcircled{P}$$

FIGURE 3.2 A schematic version of one type of enzyme catalysis. The overall reaction catalyzed by this hypothetical enzyme is the transfer of a phosphate from A to B. First, A–P and B are bound to a specific enzyme because of its surface geometry. In the complex that is formed, the phosphate comes into close contact with B, increasing the chance for the phosphate to be transferred. If no enzyme was present, the chance that A–P and B would collide in such a manner that the phosphate could be transfered would be greatly reduced. After the complex has dissociated to form A and B–P, the enzyme is released and can be reutilized. Such catalyzed reactions can occur at fantastic rates, greater than a million per minute.

catalyzed by a specific enzyme. In the more general sense it is the application of chemical principles to the solution of biological problems that best describes the discipline of *biochemistry*.

3·4 The flow of energy in the biological world

With the principles of thermodynamics and enzyme catalysis as a foundation, we are now ready to discuss energy transformations in biological systems. For purposes of discussion, cellular metabolism can be split into two group-

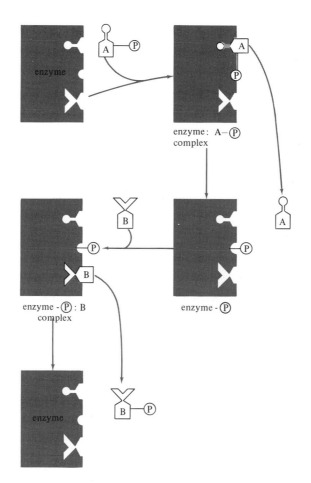

enzyme

enzyme: A—Ⓟ
complex

enzyme -Ⓟ : B
complex

enzyme -Ⓟ

enzyme

FIGURE 3.3 A schematic version of a different type of enzyme catalysis. The overall reaction is the same as in Figure 3.2. However, in this case the phosphate is first transferred from A to the enzyme, and then from the enzyme to B.

ings, *energy-yielding* and *energy-requiring* reactions. In this section the overall balance between the production and expenditure of energy is discussed in general terms. Subsequently, we examine in greater detail the three basic processes by which cells obtain energy: *fermentation*, *respiration*, and *photosynthesis*.

The most general energy-requiring process is termed *biosynthesis*, the series of reactions by which cells build large and complex molecules from small and relatively simple components. In Chapters 5 and 6 the biosynthesis of nucleic acids and protein is discussed in detail. Another energy-requiring process, *osmotic work*, is necessary for the transport and concentration of specific nutrients from the environment into the cell. Since molecules tend to randomize (second law of thermodynamics), it takes energy to concentrate the nutrients in the cell and pump the waste materials out. Many cells are able to concentrate chemicals in such a manner that the inside of the cell has a different electric charge from the outside fluid. The difference in charge between the cell and its surroundings makes possible *electric work*, the conduction of nerve impulses. A fascinating example of electric work is seen in the electric eel, which can deliver a shock of over 300 volts. *Motion* of some sort is present in all cells, whether it is the crawling of ameba, the swimming of sperm cells, the contraction of muscle cells, or the movement of chromosomes during mitosis. In addition to these four general energy-requiring processes some organisms need energy for specialized tasks, such as the production of light by fireflies and glow worms (bioluminescence).

For this general discussion, the two types of energy metabolism can be represented as follows:

$$A \xrightarrow[\text{yielding}]{\text{Energy}} B \quad \text{and} \quad C \xrightarrow[\text{requiring}]{\text{Energy}} D$$

In the energy-yielding reaction, A is converted into a product of lower energy, B. Since the total energy of the system must be conserved during the process, the difference in energy between A and B has to be released in some other form. For example, when you digest a beefsteak, the complex protein molecules are broken down into molecules of lower energy content, carbon dioxide and water; during the process part of the energy is released as heat, the remainder is utilized to promote energy-requiring processes. Conversely, the energy-requiring reaction C → D, cannot proceed unless provided with additional energy. As it stands the reaction is thermodynamically impossible because D is at a higher energy level than C. For D to be formed energy must somehow be supplied. Since many energy-requiring processes are essential for life, a basic problem of cells is: How can the energy produced by energy-yielding reactions be used to "push" energy-requiring processes?

During the course of evolution living organisms have developed an efficient method for linking reactions which consume energy and those which supply it. The method (Figure 3.4) utilizes as an intermediate "bridge" or linking system the interconversion of the all-important molecules *adenosine triphosphate* (ATP) and *adenosine diphosphate* (ADP).

The ATP:ADP system is similar to a battery. Forming ATP is equivalent to charging the battery and requires energy. When ATP is broken down to ADP (discharging), the released energy can be used to do work. In living cells the chemical linkage system works as shown in the following:

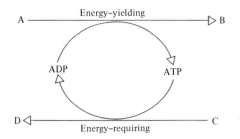

The energy released during the conversion of A to B is utilized to form ATP (charging); the energy needed for the conversion of C to D is supplied by the breakdown of ATP (discharging). In this way energy-yielding and energy-requiring processes are intimately coupled.

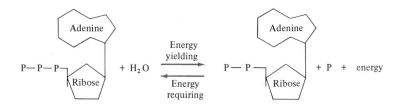

FIGURE 3.4 The interconversion of adenosine triphosphate (ATP) and adenosine diphosphate (ADP). The ATP molecule is composed of a base (adenine) joined to a sugar (ribose) which is also connected to three phosphate groups, abbreviated P. The ADP molecule has an identical structure except that it contains one less phosphate. Considerable energy is liberated when ATP and water react to form ADP and phosphate; conversely, energy is required for the reverse reaction, the synthesis of ATP.

In the economy of the cell, ATP has been compared to money in that it can be made and spent in a variety of ways. Adenosine triphosphate produced in one part of the cell can be stored and used anywhere in the cell whenever the need arises. Thus ATP plays the primary role of "middleman" between energy-producing and energy-requiring processes in all living organisms, whether they be microbes, animals, or plants.

The remaining sections of this chapter are devoted primarily to a more detailed discussion of the different ways in which cells produce ATP. In Chapters 5, 6, and 7 we shall see how ATP is utilized as the immediate source of energy for cellular synthesis and growth.

3•5 Fermentation: "La vie sans air"

Fermentation is a relatively simple and primitive biochemical process by which certain living cells are able to obtain energy in the absence of air.

From a general thermodynamic point of view the process can be stated as follows: Organic molecules are broken down into simpler substances of lower energy content; some of the free energy that is liberated during the process is captured in the formation of ATP. For example, when yeast transform a molecule of glucose into ethyl alcohol and carbon dioxide, two ATP molecules are formed concurrently.

Following Buchner's discovery of a cell-free extract that could catalyze fermentation, it became possible to study the various intermediate steps of the process in great detail. By the late 1930s, owing to the experiments of a number of eminent biochemists, including the Englishman Arthur Harden and the German Otto F. Meyerhof, the exact fate of the glucose molecule during its degradation was established. For this discussion, however, we need to deliberate only on the essential features shown in Figure 3.5.

The fermentation of glucose can be considered to occur in two stages. In the first stage, one molecule of the sugar is broken down into two molecules of pyruvic acid and the equivalent of four hydrogen atoms.[6] It should be pointed out that the formation of pyruvic acid from glucose actually entails at least ten distinct and sequential reactions. Each of these reactions is catalyzed by a specific enzyme. The enzymes work in tandem to produce two pyruvic acid and two ATP molecules for each glucose molecule consumed. Although there may be differences in detail, the overall reaction which takes place in the first stage of fermentation is a common theme of metabolism.

Fundamental differences in energy metabolism arise as a consequence of how the pyruvic acid and hydrogen are further metabolized. As we discuss

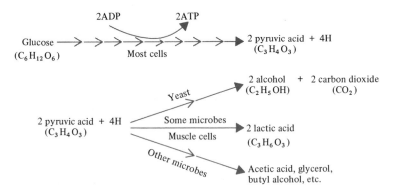

FIGURE 3.5 An outline of the sequence of reactions for the fermentation of glucose.

[6] For simplicity we will refer to hydrogen atoms in the remaining sections of this chapter. However, it should be realized that hydrogen atoms never actually exist free in the cell, but are transported by specific carrier molecules.

in the next section, cells which are able to carry out the process of respiration burn pyruvic acid all the way to carbon dioxide and water; the hydrogen atoms react indirectly with oxygen gas to form water. In the absence of oxygen, cells are faced with the serious problem of how to dispose of the hydrogen atoms. Essentially, the second stage of fermentation is the addition of hydrogen to pyruvic acid in one or more steps. The different fermentation products are a result of slight variations in this second stage. For example, in the simplest case, hydrogen is added directly to pyruvic acid to produce lactic acid. Yeast, on the other hand, first split off carbon dioxide from the pyruvic acid and then add the hydrogen to form ethyl alcohol.

Whereas only a few types of microorganisms derive energy exclusively by fermentation, many types have the capacity to perform either fermentation or respiration. These *facultative* organisms carry out respiration as long as oxygen is available to them. When the oxygen supply is depleted, they switch to fermentation. This optional behavior is by no means limited to microorganisms. Certain tissues of plants and animals are also able to carry out fermentation when deprived of oxygen. A well-studied example is muscle tissue. Muscle contraction requires the expenditure of large quantities of ATP. During periods of vigorous exercise muscle cells utilize oxygen much faster than it can be supplied by the heart and lungs. As the cells become starved of oxygen, they switch over to fermentation and produce ATP by the conversion of glucose to lactic acid. The fact that precisely the same steps are involved in lactic acid fermentation in a human muscle cell and in a lactic acid bacterium is another example of *unity in biology*.

Let us examine briefly some alcoholic fermentation processes of economic importance from the viewpoint of the biochemical pathways we have discussed. Wine production is basically the conversion of the sugar in fruit juices (15 to 25 percent), especially grape, into ethyl alcohol (up to 14 percent) and carbon dioxide. The yeast do not have to be added since the surface of a single grape may contain 100,000 yeast cells. Conditions of oxygen deprivation are maintained in the wine-making process, so that yeast cells in order to grow must obtain their ATP by fermentation, the waste products of the process being alcohol and carbon dioxide gas. In sparkling wines, such as champagne, some of the carbon dioxide gas is captured in the bottle.

Beer and wine production differ fundamentally in that the raw material for beer is the starch found in various grains. Since yeast cannot decompose starch, the polysaccharide must first be split to the monosaccharide, glucose, before fermentation can begin. It is interesting to compare how different cultures have solved this problem. In Europe, where barley is the raw material, starch is split by enzymes released when the germinated seeds are mashed. In the Orient, where fermented beverages such as sake are made from rice, starch is split by a mold *Aspergillus*, which is present during the fermentation. The American Indian developed still a different method of preparing grain for fermentation; corn was chewed, and the mixture of ground corn and saliva was spit into a bucket, where enzymes in saliva act on the starch, thus liberating glucose. Alcoholic fermentation is also an essential step in the production of bread. In this case, however, the carbon dioxide is the important fermentation product, causing the raising of the bread. The alcohol produced is evaporated off during the baking process.

3·6 Respiration

Respiration is the major means of ATP production in animals and most microorganisms. For our purposes[7] respiration can be defined as an energy-yielding metabolic process in which foodstuff reacts (indirectly) with oxygen and is completely broken down to carbon dioxide and water with the concurrent formation of ATP.

$$C_6H_{12}O_6 + 6\ O_2 \longrightarrow 6\ CO_2 + 6\ H_2O$$

The greater efficiency of respiration (38 ATP per glucose) as compared to fermentation (2 ATP per glucose) is a result of the more complete breakdown of the sugar. When glucose is fermented to alcohol, for example, only a part of the energy stored in the sugar is released. This statement can be appreciated if we recall that considerable additional energy is liberated when alcohol is burned (oxidized) in the presence of air. The large number of steps involved in respiration make it easier to control the rate of burning, so that energy can be efficiently captured in a form, ATP, that is utilizable by the cell.

As previously discussed, the conversion of sugar to pyruvic acid is common to both fermentation and respiration. In respiration, however, pyruvic acid is further metabolized to carbon dioxide and water in a series of reactions called the *Krebs cycle* or *citric acid cycle* (Figure 3.6). This cycle was formulated in the early 1940s from biochemical data obtained primarily by two refugees of the Hitler regime, Hungarian-born Albert Szent-Györgyi[8] and German-born Hans A. Krebs.

---

[7] We will not consider here a small, but interesting, group of bacteria which can utilize inorganic materials, such as iron and sulfur, as foodstuff for respiration. Also, we will not discuss certain bacteria which use sulfate or nitrate instead of oxygen for respiration.

[8] Szent-Györgyi is one of many distinguished biochemists originally trained in medicine. His first research paper dealt with hemorrhoids. Later he went into physiology, then biochemistry, and finally, physical chemistry. Reflecting on his scientific career, Szent-Györgyi recently stated he has only one regret, "I started science on the wrong end."

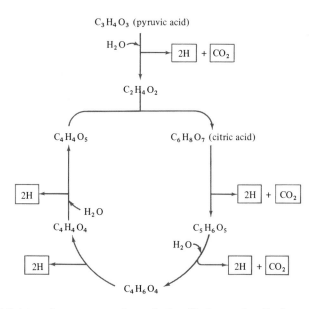

FIGURE 3.6 Schematic representation of the Krebs cycle. Each reaction is catalyzed by a separate enzyme.

Before entering the cycle, pyruvic acid is converted into a two-carbon molecule with the release of carbon dioxide and a pair of hydrogen atoms. This two-carbon molecule combines with another cellular product which contains four carbon atoms, forming a six-carbon product, citric acid. Citric acid is then broken down in a series of reactions with the stepwise release of four pairs of hydrogen atoms and two carbon dioxide molecules. Finally, the original four-carbon compound is regenerated, which can then again combine with the two-carbon molecule and start the cycle all over.

The two principal benefits that the cell derives from the Krebs cycle are the creation of *building blocks* and production of *energy*. Some of the compounds formed during this cyclic system can be used to manufacture amino acids and other essential cellular components. For example, the amino acid aspartic acid ($C_4H_7O_4N$) is made by the enzyme-controlled addition of ammonia (NH_3) to the $C_4H_4O_4$ intermediate compound. In general, the Krebs cycle serves as the hub of the cell. Intermediate products can be siphoned off at different points in the cycle and utilized to form the variety of building blocks needed for biosynthesis; conversely, molecules which are not needed as building blocks can be fed into the cycle at different places, thus allowing the cell a greater diversity of usable foodstuffs.

The Krebs cycle explains how pyruvic acid is broken down to carbon dioxide, but it does not explain why oxygen is necessary for respiration nor how ATP is produced. The specific aspect of respiration that is concerned with how oxygen utilization is coupled to ATP formation is called *oxidative phosphorylation.*

Respiration, including both the Krebs cycle and oxidative phosphorylation, takes place in highly organized subcellular bodies, the *mitochondria* (see Figure 2.20). Almost all animal and plant cells contain numerous mitochondria in their cytoplasm. The detailed structure of the mitochondrion, as revealed by the electron microscope, is discussed in Chapter 2. It is now possible to separate mitochondria from other cell components. A suspension of these purified mitochondria under appropriate conditions is able to convert pyruvic acid to carbon dioxide and water with the formation of ATP. Since mitochondria can perform all of the reactions involved in respiration, it follows that all of the enzymes and other components necessary for the Krebs cycle and oxidative phosphorylation must be located in mitochondria. *The mitochondria thus serve as self-contained powerhouses for the cell.*

Utilizing purified mitochondria, oxidative phosphorylation has been investigated during the last 25 years by a number of distinguished biochemists, including Britton Chance at the University of Pennsylvania, Albert L. Lehninger at Johns Hopkins University, and David E. Green at the University of Wisconsin. Although the exact details are unknown, the general scheme can be represented as shown in Figure 3.7.

First, a pair of hydrogen atoms combine with carrier A to produce AH_2. Next, AH_2 transfers the hydrogens to carrier B, producing BH_2 and regenerating A. This energy-yielding reaction is coupled to the formation of ATP from ADP and phosphate:

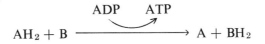

$$AH_2 + B \xrightarrow{\quad\quad\quad\quad\quad} A + BH_2$$

Carrier A again combines with a pair of hydrogen atoms to form AH_2; BH_2 transfers the hydrogens to carrier C, producing CH_2 and regenerating B. In

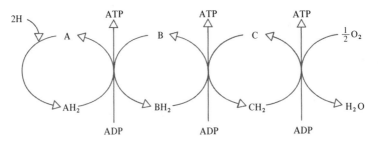

FIGURE 3.7 Schematic representation of oxidative phosphorylation. Hydrogen atoms produced during the breakdown of nutrient molecules are transmitted to oxygen by a series of carrier molecules (abbreviated A, B, and C). Three ATP molecules are formed for each pair of hydrogen atoms transferred to oxygen.

this way the hydrogens are transferred through the carrier[9] molecules, finally combining with oxygen gas to form water. *The net result of this chain of reactions in oxidative phosphorylation is the formation of 3 ATP molecules:*

$$3ADP \qquad 3ATP$$

$$2\,H + \tfrac{1}{2}\,O_2 \xrightarrow{\hspace{3cm}} H_2O$$

The importance of the oxidative phosphorylation chain to living organisms is demonstrated by the lethal effect of two well-known poisons, carbon monoxide and cyanide. Both toxins exert their effect by combining with carrier molecules used in the oxidative phosphorylation chain, thus preventing transfer of hydrogens and interrupting the chain. This results in an immediate halt in ATP production, and unless the toxin is removed rapidly, death ensues.

We can now compute the total number of ATP molecules formed for each molecule of glucose consumed. In the first phase, the conversion of glucose to pyruvic acid, two ATP and two pairs of hydrogen atoms are formed. When the two pyruvic acid molecules are burned to cabon dioxide via the Krebs cycle (Figure 3.5), ten pairs of hydrogens are produced, making a total of two ATP and twelve pairs of hydrogen atoms. Since each pair of hydrogens yields three ATP, a grand total of *38 ATP* molecules are manufactured for each glucose that is oxidized to carbon dioxide and water.

It is interesting to relate the ATP yield to the caloric value of food. For example, 7 grams of glucose (1 teaspoonful) contains about 2.5×10^{22} molecules. Since each glucose molecule gives rise to 38 ATP, $38 \times 2.5 \times 10^{22}$, or 9.5×10^{23} molecules of ATP are produced for each teaspoonful of glucose consumed. Biochemists have determined that the conversion of one ATP molecule to ADP releases 1.6×10^{-23} Calories[10] of useful energy. Thus a teaspoonful of sugar contains $1.6 \times 10^{-23} \times 9.5 \times 10^{23}$, or approximately 15 Calories.

3·7 Photosynthesis: The ultimate source of energy

As you probably already know, the ultimate source of energy for most living organisms on earth is the sun. A continuous series of nuclear reactions take place on the sun, releasing vast quantities of energy. Part of this solar energy

---

[9] Actually, at least seven different carrier molecules are involved in transferring the hydrogens. However, in only three of these transfers are ATP produced.
[10] The Calorie used in nutrition is defined as the amount of heat energy necessary to raise one liter (about a quart) of ice-cold water 1°C.

is in the form of visible light, and *photosynthesis* is the process by which plants (and certain microorganisms) utilize light as the source of energy for the production of carbohydrate.

Early history

Before discussing modern concepts on the mechanism of photosynthesis, let us sketch briefly some of the early history. Probably the first important study of photosynthesis was published in 1648 by the Dutch physician Jean-Baptiste van Helmont:

> I took an earthenware pot, placed in it 200 lb. of earth dried in an oven, soaked this with water, and planted in it a willow shoot weighing 5 lb. After five years had passed, the tree grown therefrom weighed 169 lb. and about 3 oz. But the earthenware pot was constantly wet only with rain water; . . . . Finally, I again dried the earth of the pot, and it was found to be the same 200 lb. minus about 2 oz. Therefore, 164 lb. of wood, bark, and root had arisen from the water alone.

In this classic experiment van Helmont considered only soil and water as possible sources for the increased mass of the willow tree. Since the weight of the soil did not change appreciably, his conclusion was obvious—the tree grew at the expense of water alone. In implicating the importance of water as a raw material for plant growth, van Helmont was perfectly correct; we now know, however, that water is not the only essential ingredient.

In 1727 the English physiologist Stephen Hales published the results of an experiment that demonstrated for the first time that air as well as water was necessary for plant growth. Hales added ample quantities of water to two containers full of soil; in one of the containers he set a small peppermint plant; the other (the control) contained only soil and water. Both containers were then made air-tight by placing inverted glass jars over them. The mint grew for about nine months, then faded and died. Next, Hales exchanged the dead mint for a fresh plant utilizing a technique which prevented fresh air from entering the container during this transfer. This time the plant died in four or five days. When a fresh plant was placed, using the same technique, in the control container whose air had also been confined for nine months, the plant lived several months. Hales concluded from these experiments that plants interact with the atmosphere, possibly removing a substance essential for growth. The combined data of van Helmont and Hales can then be summarized as

$$\text{Water} + \text{air} \rightarrow \text{Plant material}$$

The next important development was the discovery in 1772 by Joseph Priestly that green plants, "instead of affecting the air in the same manner as animal respiration, reverse the effects of breathing, and tend to keep the atmosphere sweet and wholesome." One of Priestley's experiments is summarized in Figure 3.8.

It was already known that fresh air was necessary for both animal and plant life. Preistley's contribution was the discovery that plants and animals grown in the same atmosphere are mutually beneficial: Each restores something tc the air that the other consumes. Plant growth could then be expressed as

$$\text{Water} + \text{gas X} \rightarrow \text{Plant material} + \text{gas Y}$$

with the added stipulation that X and Y are the same gases that are produced and consumed, respectively, during animal respiration.

Priestley's results were soon elaborated on by a Dutch physician named Jan Ingen-Houz. Ingen-Houz demonstrated that the production of gas Y occurred *only in the green parts of plants and only in the light.*

> All plants posses a power of correcting, in a few hours, foul air unfit for respiration, but only in clear daylight, or in the sunshine . . . This office is not performed by the whole plant, but only by the leaves and the green stalks that support them.

Since the production of plant material and gas Y required light, the process came to be called "photosynthesis," which means "put together by light." By the end of the eighteenth century, photosynthesis could be expressed by the following equation:

$$\text{Water} + \text{gas X} \xrightarrow[\text{Green cells}]{\text{Light}} \text{Plant material} + \text{gas Y}$$

With subsequent developments in chemistry, the increased weight of plants during photosynthesis was shown to be due to the formation of glu-

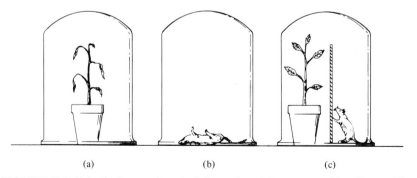

(a)   (b)   (c)

FIGURE 3.8 Priestley's experiment. A lone plant (a) or mouse (b) suffocated in a closed jar. When a plant and mouse were enclosed in the same jar (c), both survived.

cose, gas X was carbon dioxide and gas Y was oxygen. Thus the overall photosynthetic process could be expressed in more precise chemical terms:

$$6 \ H_2O + 6 \ CO_2 \xrightarrow[\text{Green cells}]{\text{Light}} C_6H_{12}O_6 + 6 \ O_2$$

Modern view

Early investigations of photosynthesis were primarily concerned with identifying each of the components involved in the process. As you probably have perceived from the equation above, the net result of photosynthesis is the reverse of respiration. Since respiration is an energy-yielding process, it follows that the formation of sugar during photosynthesis is an energy-requiring process. In recent years research into the photosynthetic process has focused on the problem of *how* light energy is trapped and utilized to "drive" the synthesis of sugar from carbon dioxide and water.

Important clues regarding the mechanism of photosynthesis came from studies of two specialized groups of bacteria. One group, termed *chemosynthetic bacteria*, obtains its energy (ATP) by burning (oxidizing) inorganic compounds. The other group obtains energy by performing an "abnormal" type of photosynthesis. In both cases the energy is utilized to convert carbon dioxide from the atmosphere into cellular material.

In 1929 Cornelius B. van Niel of Stanford University proposed a hypothesis which applied not only to green plant photosynthesis but also to bacterial photosynthesis and chemosynthesis. It is truly enlightening to see how van Niel arrived at his unifying hypothesis from some rather simple analytical data.

The results of van Niel's quantitative experiments with the red photosynthetic bacteria can be expressed by the following equation:

$$12 \ H_2S + 6 \ CO_2 \xrightarrow[\text{Red bacteria}]{\text{Light}} C_6H_{12}O_6 + 6 \ H_2O + 12 \ S$$

This "abnormal" photosynthesis is similar to plant photosynthesis except that the bacteria use H_2S (hydrogen sulfide) instead of H_2O and produce sulfur instead of oxygen. The comparison is even more striking if six water molecules are added to each side of the equation for plant photosynthesis:

$$12 \ H_2O + 6 \ CO_2 \xrightarrow[\text{Green plants}]{\text{Light}} C_6H_{12}O_6 + 6 \ H_2O + 6 \ O_2$$

Combining the data from bacterial and plant photosynthesis and dividing by six, van Niel was able to express a general equation for photosynthesis:

$$2 H_2A + CO_2 \xrightarrow[\substack{\text{All photosynthetic} \\ \text{cells}}]{\text{Light}} (CH_2O) + H_2O + 2 A$$

where A stands for either oxygen or sulfur and (CH_2O) symbolizes cell matter having a ratio of two hydrogens for each carbon and oxygen.

Since in bacterial photosynthesis the sulfur can only come from H_2S, van Niel argued by analogy that the *oxygen produced during photosynthetis arose by splitting H_2O*. (This part of van Niel's hypothesis was experimentally verified in the early 1940s by Samuel Ruben and his collaborators.) The generalized equation for photosynthesis can thus be considered as the sum of two processes:

$$(1) \qquad 2 H_2A \xrightarrow{\text{Light}} 4 H + 2 A$$

$$(2) \qquad 4 H + CO_2 \xrightarrow[\text{or dark}]{\text{Light}} (CH_2O) + H_2O$$

$$(1) \text{ plus } (2) \qquad 2 H_2A + CO_2 \longrightarrow (CH_2O) + H_2O + 2 A$$

By writing photosynthesis as a two-step process, it is possible to derive an even more general concept of cell metabolism. Although reaction (1) occurs only in the light and is peculiar to photosynthetic organisms, reaction (2) is common to a large variety of cells and can proceed in the dark. In essence, the first reaction is the *conversion of light energy into chemical energy*. The energy is then able to "drive" the second reaction, the formation of cell material from carbon dioxide. The significance of separating the "photo" reaction from the "synthesis" reaction is that it now brings into the framework of this discussion a number of other organisms. For example, an interesting group of bacteria obtain their energy by oxidizing inorganic substances, such as iron or sulfur. The hydrogen atoms and energy so derived are then utilized to convert carbon dioxide to cellular material exactly as in the second reaction of photosynthesis. In the most general sense, the formation of organic matter from carbon dioxide was expressed by van Niel as follows:

$$CO_2 \xrightarrow{4 H} (CH_2O) + H_2O$$

where the hydrogens can be derived in a variety of ways, such as "photo" reactions or the burning of inorganic compounds.

It is interesting to compare the development of van Niel's conceptual scheme with the progressive abstraction that has taken place in modern art.

The exquisite series of bas-reliefs of Henri Matisse will serve to illustrate the point (Figure 3.9). Both Matisse and van Niel begin by examining in detail

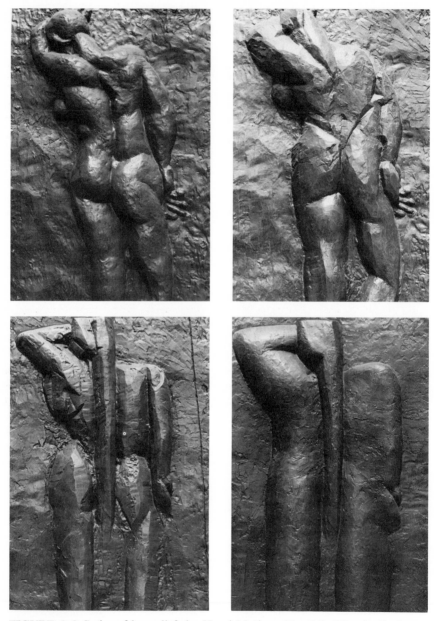

FIGURE 3.9 Series of bas-reliefs by Henri Matisse. (Franklin Murphy Sculpture Garden, University of California at Los Angeles.)

a specific case—for van Niel, the red photosynthetic bacteria; for Matisse, the rear view of a beautiful woman. Both creators, scientist and artist, attempt to abstract from reality the essence of their subject, slowly constructing a greater simplicity and universality. But as in art, so in science, the final form is forever tentative.

In the last 25 years further progress has been made in elucidating the mechanisms of both the "photo" or *light reaction* and "synthetic" or *dark reaction*. From the experiments of Melvin Calvin and his associates at the Lawrence Radiation Laboratory of the University of California, the dark reactions of photosynthesis are now known in some detail. As shown in Figure 3.10, carbon dioxide does not react directly with hydrogen atoms to form organic molecules; instead carbon dioxide is converted into an organic substance by combining with a five-carbon sugar diphosphate to produce a six-carbon sugar acid. It is not our intention to delve into the details of this important metabolic pathway. The crucial points are that ATP is necessary to form the five-carbon acceptor molecule and that *hydrogen* (reducing power) is necessary to convert the six-carbon acid to a six-carbon sugar. Adenosine triphosphate and hydrogen are thus necessary for the formation of carbohydrate from carbon dioxide even though they are not utilized in the specific reaction in which carbon dioxide is fixed.

Many different types of cells, including nonphotosynthetic bacteria, are able to convert carbon dioxide to sugar *if provided reducing power* (*hydrogens*) *and ATP*. This implies that the function of the light reaction is to generate energy in the form of both hydrogen and ATP. As we have discussed, the hydrogens come from the splitting of water. But how is ATP formed?

At first it was thought that the ATP needed for fixing carbon dioxide in photosynthesis is produced in mitochondria by oxidative phosphorylation. According to this hypothesis, some of the hydrogens generated by the light

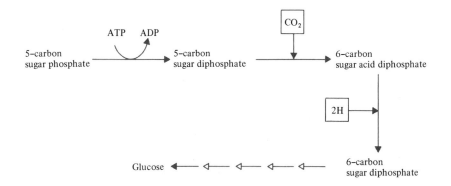

FIGURE 3.10 Greatly simplified representation of the dark reactions of photosynthesis.

reaction would function exactly like the hydrogens formed during the Kreb's cycle.

It has been known for more than 100 years that photosynthetic eucaryotic cells contain in their cytoplasm specialized structures called *chloroplasts* (see Figs. 2.23 and 2.24). In 1954 Daniel I. Arnon made the important discovery that *isolated chloroplasts can carry out the complete photosynthetic process*. It follows that all of the components necessary for producing ATP and hydrogen from light energy as well as the enzymes needed for the dark reactions must be present within the chloroplast. Recently, Arnon has provided direct experimental evidence for the generation of ATP in chloroplasts. This light dependent, energy-yielding reaction is called *photosynthetic phosphorylation* to distinguish it from oxidative phosphorylation.

Figure 3.11 is a schematic representation of one of the ways in which Arnon has suggested that ATP is formed in the chloroplast. The chloroplast contains the green pigment *chlorophyll* which traps solar radiant energy. The light energy impels an electron into a higher energy state, and the chlorophyll molecule acquires, temporarily, a positive charge. The high energy electron then passes through a series of carrier molecules, A, B, C, and D, finally relapsing to its original energy state and completing the cycle. The loss in energy as the electron drops from A back to chlorophyll is conserved in the formation of two ATP molecules. Although the physical details of this series of reactions is not completely understood, the importance of chlorophyll now becomes clearer: Through this molecule passes energy from the sun, that allows plants to grow, animals to move, and men to think.

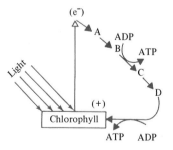

FIGURE 3.11 Schematic representation of photosynthetic phosphorylation.

THOUGHT AND REVIEW QUESTIONS

3.1 Define entropy. State the second law of thermodynamics in terms of entropy.

3.2 We all know that living organisms are able to maintain and reproduce what appears to be a high degree of complexity and organization out of relatively random surroundings. Does this mean that living organisms do not obey the second law of thermodynamics? Explain. How about the construction of a building from bricks?

3.3 Which of the energy-yielding processes, fermentation, respiration, or photo-synthesis, do you think evolved first on earth? Which was second?

3.4 Estimate how many different enzymes are required for the conversion of glucose to carbon dioxide with the formation of 38 ATP molecules.

3.5 Which of the following reactions are energy yielding? Which require ATP?

(a) Pyruvic acid + oxygen $\rightarrow$ Carbon dioxide + water
(b) Amino acids $\rightarrow$ Protein + water
(c) $C_4H_6O_4 \rightarrow C_4H_4O_4$ + 2 H atoms
(d) Starch + water $\rightarrow$ Glucose
(e) Carbon dioxide + water $\rightarrow$ Oxygen + plant material

3.6 In terms of their nutritional requirements, explain why, in general, plants have a large surface area (flat leaves, and so on) and are not motile, whereas animals are compact and motile.

3.7 From the viewpoint of an animal, what is the importance of photosynthesis?

3.8 Explain why cells which obtain their energy by respiration require less sugar than cells performing fermentation.

3.9 Distinguish between the "light reactions" and "dark reactions" of photo-synthesis.

3.10 Biology can be viewed in terms of unity and diversity. Which of the following statements apply to all living organisms and which are true of only certain organisms?

(a) Produce their own ATP
(b) Carry out respiration
(c) Are motile
(d) Have nuclear membranes
(e) Are destroyed at 120°C for 20 minutes
(f) Depend on other organisms for organic nutrients
(g) Depend on other organisms for either organic or inorganic molecules
(h) Have cytoplasmic membranes
(i) Require oxygen for growth
(j) Contain a large number of enzymes

SUGGESTED READINGS

Amerine, M. A., "Wine," *Scientific American*, August 1964 (offprint 190). A popular account of the wine-making process.

Asimov, I., *Life and Energy*. New York: Bantam Books, Inc., 1962. An introduction into the chemistry of energy processes by one of the world's foremost science writers.

Gabriel, M. L., and S. Fogel (eds. and trans.), *Great Experiments in Biology*. Englewood Cliffs, N.J.: Prentice-Hall, Inc., 1955. An excellent collection of important papers, including the classical experiments of Buchner, Pasteur, van Helmont, Priestly, Ingen-Houz, and van Niel.

Lehninger, A. L., *Bioenergetics*. New York: W. A. Benjamin, Inc., 1965. A more advanced treatment of many aspects of this chapter.

Levine, R. P., "The Mechanism of Photosynthesis," *Scientific American*, December 1969.

Nash, L. K., "Plants and the Atmosphere," *Harvard Case Histories in Experimental Science*, Vol. II, Cambridge, Mass.: Harvard University Press, 1957. A historical account of photosynthesis.

4

GENETICS: GARDEN PEAS TO BACTERIOPHAGES*

For eon after eon, creature has given rise to creature upon this earth — blindly, each generation usually like the former, occasionally — by accident — a little different. Of all the creatures that have lived upon this earth we are the first to understand this process. . . . The ultimate significance of this understanding of the very basis of heredity is incalculable. It will change all the eons to come.

ROBERT L. SINSHEIMER

As discussed in Chapter 1, by the end of the nineteenth century it had been firmly established that, under present conditions, life as we know it does not arise from nonliving materials. Stated in another way, it had been demonstrated that all living organisms have parents. By this time one of the great biological generalizations, commonly referred to as the cell doctrine, had been formulated. It may be stated as, "All living things are composed of cells that arise by a series of orderly divisions from preexisting cells." Thus the cell was considered to be the structural and functional basis of all living things, including the unicellular protozoa and bacteria as well as multicellular plants and animals.

Another great biological generalization, Darwin's theory of evolution, had also been generally accepted by the scientific community by the end of the nineteenth century. According to this theory the various species of plants and animals presently inhabiting the earth arose during times past by a series of small changes from ancestors that differed greatly from themselves. Thus man and monkey may have descended from a common ancestor, which, in turn, had evolved from yet a simpler kind of animal. This theory was consistent with a great many observations, including fossil records and observed functional and morphological similarities among different species of plants and animals.

Darwin's theory did not, however, explain adequately the method whereby variations from one generation to the succeeding ones arose. In his

* This chapter was written by Dr. W. R. Romig, Department of Bacteriology, University of California, Los Angeles, California.

time it was rather widely assumed that living things could change their shape and bodily functions to meet better the stresses imposed on them by environmental conditions. These environmentally imposed changes were then thought to be transmitted to their offspring so that gradually a new, better adapted species evolved. This idea is usually associated with the French zoologist Jean-Bapiste Lamarck, who strongly supported it.

Darwin's later views were that naturally occurring variants in a species arose by chance and that these variations were inherited by the offspring of parents possessing them. It was further assumed that these chance variations might be "good" or "bad" traits, so that at any particular time a species might contain members that are well or poorly fitted for a particular environment. Darwin reasoned that the driving force of evolution was the selection of the fitter types by environmental conditions, with the corresponding elimination of the less fit individuals from the population.

It was not, however, until Mendel's researches were independently rediscovered by three different investigators early in the twentieth century that a clear understanding of inheritance began to unfold. Gregor Mendel's experiments and his interpretation of them were first published in 1866, but his work was largely ignored in his lifetime and was then forgotten until rediscovered in 1900.

4•1   The experiments of Gregor Mendel

Other plant breeders before Mendel had investigated the nature of inheritance by methods similar to his. Many of these investigations were inconclusive because the inherited characteristics studied were not clearly distinguishable in the offspring. Mendel's work with the common garden pea was successful for many reasons, not least of which was his decision to concentrate on certain clearly distinguishable minor traits in the strains of peas he chose for his experimental material. Mendel used varieties of peas that showed sharply contrasting characteristics. For example, one of his plants, when crossed with itself for many generations, always produced plants with round, full seeds, whereas another variety tested under these conditions always produced wrinkled, irregularly shaped seeds; another variety produced yellow seeds, in contrast to the green seeds produced by another. Such stocks of plants are said to "breed true."

It is important to realize, however, that most of the characteristics of these plants were the same and that they differed usually in only one or a few of their properties. Thus when two different varieties of peas were bred together, Mendel was able to determine unambiguously which of the two characteristics was inherited by their offspring. Such offspring are called *hybrids* to denote that they resulted from the mating between two genetically unlike parents. You will note that this use of the term differs from the way we generally use it. Hybrid, which comes from the Latin name for the offspring of a tame sow a wild boar, is commonly the name given to the offspring of two animals or plants of different races, varieties, or species, such as the mules obtained by crossing horses and asses. Here we mean the offspring of parents that are genetically unlike in *any* characteristic.

Mendel's selection of pea plants was not made just because they were shown to breed true for the chosen characteristics. He also demonstrated that they could be artificially cross-fertilized by a method in which the pollen from one plant was collected and used to fertilize the desired maternal plant; alternatively, he found that pea plants were very efficiently self-fertilized under normal conditions with very little risk of impregnation by foreign pollen. Under these latter conditions, both the male sex cells (or *gametes*) and the female gametes with which they fuse to form new individuals are made by a single plant. Finally, Mendel realized the importance of examining large enough numbers of offspring so that his conclusions were statistically significant.

Mendel's basic experiments consisted of first artificially crossing two plants which exhibited different, distinct characteristics. The seeds resulting from such a cross were carefully gathered and planted the following spring. After these hybrid plants matured, they were examined to see which, if either, of the parental characteristics they expressed. The hybrids were then allowed to self-pollinate, and their offspring were examined for the presence or absence of the characteristics under consideration. This latter procedure of examining the progeny of self-pollinated hybrids could be continued indefinitely, and Mendel studied as many as six successive generations of plants in this manner. It is convenient to keep track of the different generations in a systematic way, and for this purpose the *parental generation* is symbolized as P_1; the resulting *hybrid generation* is referred to as the F_1 (first filial) generation; the progeny resulting when F_1 individuals are crossed to each other are termed the F_2 generation, and so on.

One might predict that if two plants were crossed, one having yellow seeds and the other having green seeds, the hybrids (F_1) would possess seeds intermediate between the two. However, in Mendel's first experiments, he observed that the F_1 plants from such a cross had only yellow seeds. To explain the results of this cross, Mendel coined two terms, *recessive* and *dominant*. The yellow seed characteristic inherited from one parent is *dominant* to the green seed characteristic inherited from the other. The green seed characteristic in this example is *recessive* to the yellow.

These results raised several questions. Had the potential for producing green seeds simply been excluded from the seeds from which the hybrids were grown? Were both characteristics transmitted by the parents to the hybrids, but the green color masked in some manner by the yellow color inherited from the other parent? Had some sort of blending of the characteristics occurred in which the yellow color emerged in a greater preponderence?

To answer these questions, Mendel allowed the individual plants comprising the hybrid F_1 generation to self-pollinate, planted the seeds they produced, and examined the resulting F_2 generation. He observed that both plants with green seeds as well as plants with yellow seeds appeared in this generation. He found that the plants with yellow seeds comprised $\frac{3}{4}$ (6022) of the entire F_2 crop (8023), whereas those with green seeds accounted for about $\frac{1}{4}$ (2001). This result, of course, conclusively demonstrated that the green seed characteristic had not been eliminated from the F_1 hybrid nor had it been "blended" with it in some manner. Instead this characteristic had been transmitted intact to the F_1 hybrids, where its presence was "masked" in some manner by the dominant factor. Mendel further found that plants from the F_2 generation which exhibited the recessive green color continued to exhibit this characteristic only in the progeny (F_3) issuing from further self-fertilizations; in other words, they bred true. This was *not* the case when plants producing the dominant yellow seed color were allowed to reproduce further by self-fertilization. *Approximately* $\frac{2}{3}$ of them (60 of 100) produced offspring that displayed either the dominant or recessive character in the proportion of 3 yellow to 1 green just like the F_1 generation; the remaining $\frac{1}{3}$ (40 of a total of 100) bred true, always exhibiting the dominant yellow characteristic in further generations.

Mendel reported most of his results at a meeting of the Brünn Society for the Study of Natural Science in 1865. Although his paper was received with polite attention, it was recorded in the minutes that "there were neither questions nor discussion." Apparently the fact that scientific history was made that night went unnoticed. It should be pointed out that in his classic paper Mendel not only demonstrated the simple mathematical relationship involved in inheritance but also correctly interpreted his results by making a few reasonable assumptions. However, rather than giving his interpretation directly, it is more convenient to explain Mendel's observations using modern terms and concepts.

Each of the contrasting traits present in the pure breeding parents is controlled by a *pair* of identical factors called genes. We now know that genes are located on structures called chromosomes which are present in all cells. Each body cell (or *somatic* cell) of an individual contains two chromosomes bearing the genes controlling each trait. Such cells are called *diploid*. If the genes of a diploid cell are identical, as in the pure breeding plants, the individual is *homozygous* for that pair of genes or chromosomes. (It is also pertinent to note here that the total number of chromosomes per somatic cell, although constant for each cell within a given species, may vary considerably from one kind of plant or animal to another. Thus human body cells are all observed to have 23 pairs of chromosomes (Figure 2.29), whereas the peas used by Mendel contain seven pairs of chromosomes per somatic cell.)

During the formation of the sex cells (gametes), a process called *meiosis* occurs by means of which the total number of chromosomes is reduced by half, so that only one member of each set of chromosomes is included in the male or female gamete. The gametes, then, in contrast to the somatic cells, are *haploid* since they contain only one copy of the gene controlling each characteristic of an individual. Figure 4.1 illustrates how meiosis reduces the chromosomal complement to the haploid state.

As a result of fertilization, the male sex cell of one parent fuses with the female sex cell of the other parent so that the resulting fertilized egg, like the parental somatic cells, possesses two genes controlling each characteristic.

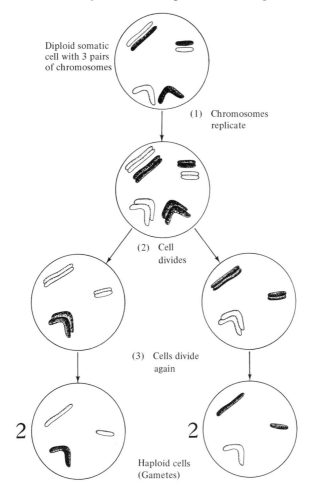

Diploid somatic cell with 3 pairs of chromosomes

(1) Chromosomes replicate

(2) Cell divides

(3) Cells divide again

Haploid cells (Gametes)

FIGURE 4.1 Highly schematic representation of the formation of haploid gametes during meiosis. These changes are brought about by a single chromosomal duplication followed by two successive nuclear divisions.

The hybrid plant develops from this fertilized egg cell by an orderly series of mitotic divisions (see Figures 2.27 and 2.28) and is *heterozygous* for the characteristic under consideration, having received a gene for yellow seeds from one parent and a gene for green seeds from the other. These two genes are called *alleles*, which is the name given to different forms of the same gene.

It should be noted, however, that it is not directly possible to determine the nature of the gene pair present in these hybrids. Because the yellow seed determinant is dominant to the green determinant, all the plants of the F_1 generation appear identical to the homozygously yellow parent. In this situation we say that the hybrids are phenotypically indistinguishable from the yellow parent. The term *phenotype* refers to the observable properties of an organism, and *genotype* refers to its genetic constitution.

It is possible to obtain pertinent information about the genotype of the F_1 hybrids by allowing them to reproduce by self-fertilization. Once again in the process of gamete formation, the two genes for each characteristic of the plant are separated, and each male and female gamete receives one of each. This results in equal numbers of male and female gametes that have received the gene for either the yellow seed or the green seed, but, of course, none will have received both. When the male gametes fertilize the female egg cells, it is strictly a matter of chance which male gamete (yellow or green) fuses with which female gamete (yellow or green). The resulting progeny should therefore contain those individuals which have received a "green" gene from both gametes, those which have received "yellow" genes from both gametes, and those which have received one of each kind from either a male or a female gamete. If the process occurs as outlined, the phenotypes of this F_2 generation should show that $\frac{3}{4}$ of the individuals have received at least one of the "yellow" determinants from either the male or female gametes (or both) and hence produce yellow seeds, and $\frac{1}{4}$ should have received a "green" determinant from both kinds of gametes and hence should express the green seed phenotype. This process is illustrated diagramatically in Figure 4.2.

Of the seven pairs of alternative traits studied by Mendel, one allele of each gene pair was always dominant to the other. Examples we have considered are yellow seeds which are dominant to green seeds and round seeds which are dominant to wrinkled seeds. Later experimenters working with other plants showed that dominance of one trait to another alternative trait is frequently not observed. Let us consider two pure breeding plants which produce either red or white flowers. We can represent the genes of the red plant as *RR* and those of the white plant by *rr*. When crossed together, the F_1 hybrids, *Rr*, are neither red nor white, but instead are all pink.

At first we might think that these results contradict those reported by Mendel, and that the "red gene" and "white gene" have blended together to produce the pink phenotype displayed by the hybrids. However, if the pink F_1 hybrids are crossed together, we find the F_2 generation contains plants of

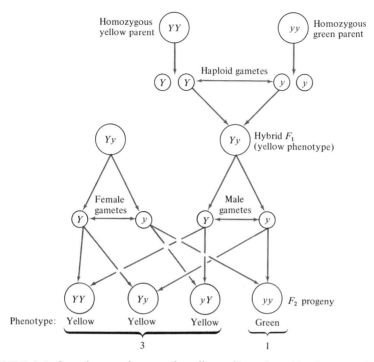

FIGURE 4.2 Genetic cross between breeding yellow plant (dominant) and green plant (recessive). Large *Y* represents gene for yellow color; small *y* represents gene for green color. The parental plants are crossed by artificial pollination, and the hybrids produced are allowed to self-pollinate. F_2 progeny show a typical 3:1 Mendelian ratio.

which ¼ are red, ¼ are white, and the remaining half are pink. These results are readily interpretable on the same basis as those in our first example. They show again that genes in the hybrid do not mix, and in future generations they separate cleanly from the allelic gene with which they were associated. In addition, this example illustrates that some genes are not dominant to one another and when both are present in a hybrid, the resulting phenotype may be intermediate between the phenotypes of the two homozygous plants. These experiments are presented in Figure 4.3.

Mendel's next series of experiments considered the consequences of breeding together two plants each differing from the other by two well-defined traits instead of only one. Let us illustrate with plants having the dominant traits of yellow seeds which are also full and round in shape, and another strain which always produces green seeds which are wrinkled and irregularly shaped.

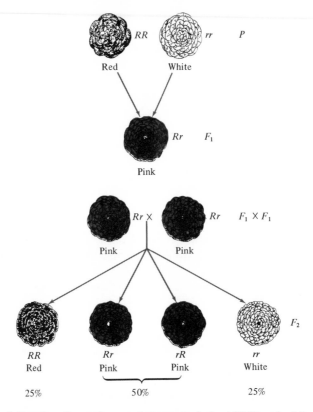

FIGURE 4.3 Results of genetic cross between red plant (*RR*) and white (*rr*) plant. Neither gene is dominant, so hybrid color is intermediate (pink). Results of crossing $F_1 \times F_1$ show both genes separate again.

These may be represented as *YYFF*, where *Y* stands for the gene determining yellow seeds and *F* stands for the gene determining the full seed shape. The other strain can be represented as *yyff*, where *y* stands for green seeds and *f* represents the gene for wrinkled seeds. When male and female gametes are formed, they will each receive *YF* from one parent and *yf* from the other. The hybrids formed from fusion of these unlike gametes will then be *YyFf*, and they will all produce the yellow, full seeds typical of the dominant parent.

When these hybrids are crossed to one another, their progeny are found in the proportions of $\frac{9}{16}$ yellow and full; $\frac{3}{16}$ yellow and wrinkled; $\frac{3}{16}$ green and full; and $\frac{1}{16}$ green and wrinkled.

It is instructive to examine the proportions of each trait in the total progeny independently of all the others. It is evident when only the yellow color is considered, these progeny represent $\frac{3}{4}$ of the total ($\frac{9}{16} + \frac{3}{16}$), whereas green color is present in $\frac{1}{4}$ of the total ($\frac{3}{16} + \frac{1}{16}$). The same proportions of full seeds, $\frac{3}{4}$, and wrinkled seeds, $\frac{1}{4}$, are also represented. Thus it is evident that the same 3:1 phenotypic ratio previously observed in the F_2 generation derived from plants hybrid for only the dominant yellow and recessive green seed colors is also present in this cross.

These results suggest that the segregation of one set of genes is not affected by the presence of another pair of different genes, and that the mechanism by which gametes are formed may be the same whether we consider one or more sets of different genes.

It is likewise instructive to consider the kinds of gametes that can be formed by the F_1 hybrids. Assume there are no restrictions governing which gene can enter the gamete, but that each gamete must receive one representative of all available genes. It is apparent that only four possible types of male and female gametes can be formed: *YF*, *Yf*, *yF*, and *yf*. If our assumption is true and the presence of one gene does not influence the likelihood of the gamete receiving any other kind of gene, these four types of male and female gametes should be formed in equal numbers. Then in a large number of fertilizations, it would be expected that every kind of female gamete would be fertilized by every kind of male gamete with approximately equal frequency. The results of such predictions are most easily described by the checkerboard method, and illustrate that the frequencies of different phenotypes found agree with those predicted from our assumptions (see Figure 4.4).

These three examples illustrate the two main principles derived from Mendel's work. The first is the principle of *independent segregation* whereby the two genes of a given set cleanly separate from one another during formation of the gametes. The second principle, *independent assortment*, is illustrated by the finding that the presence of one gene in a gamete does not affect the probability that it will receive either of the representatives of another set of genes.

4•2 Genes are on chromosomes

Soon after Mendel's work was rediscovered and its significance appreciated, intensive genetic studies were begun on a variety of plants and animals. One group of investigators headed by Thomas Hunt Morgan, then at Columbia University, began an extensive study of the genetics of the small fruit fly, *Drosophila melanogaster*.[1] These investigators soon identified and studied many different genes and although their results strongly supported Mendel's conclusions, they also clearly demonstrated that certain groups of genes do *not* segregate independently. If one considers what we now know about the physical basis of segregation, then the results of Morgan are predictable. As we have seen, the separation of the different allelic genes of a hybrid occurs

---

[1] For his pioneering genetic studies, Morgan was awarded the Nobel Prize in 1934.

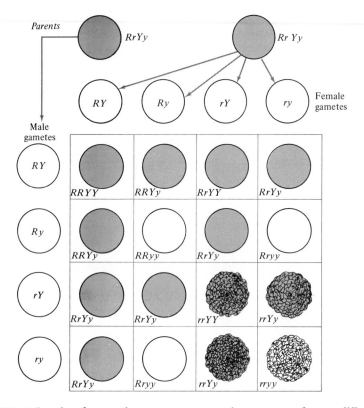

FIGURE 4.4 Results of a cross between two parents heterozygous for two different genes. The symbols are as follows: Dominant gene *R* controls round full seeds; it is dominant to its recessive allelic gene, *r*, which results in rough seeds when it is homozygous. Dominant gene *Y* controls yellow color; its recessive allele, *y*, results in green color when it is homozygous.

because the haploid gametes formed during meiosis contain only one of each kind of chromosome. Whenever two different genes are located on different chromosomes, their independent assortment is a natural consequence of meiosis. However, in *Drosophila*, which has only four pairs of chromosomes, there is a high probability of finding different genes on the same chromosome. Genes found on the same chromosome are called *linked genes*. As is expected, the inheritance of a pair of linked genes differs considerably from the inheritance of a pair of unlinked genes.

As a hypothetical example, let us consider the three linked genes, ABC, and their recessive alleles, abc, that control phenotypically distinguishable traits such as wing shape, eye color, and body pigmentation, respectively. Hybrid flies, $\frac{ABC}{abc}$, are phenotypically indistinguishable from normal flies, but their genetic makeup can be determined in a fairly straightforward manner. One way to do this utilizes the homozygously recessive parent, $\dfrac{a\ b\ c}{a\ b\ c}$, which displays an abnormal phenotype for each of the three characteristics. A *test cross* is performed in which the hybrid fly is mated (backcrossed) to the recessive parent and the resulting progeny are analyzed.

If each of these three genes were unlinked, that is, were located on a different chromosome, eight different kinds of gametes would be formed by the hybrids (A B C, A B c, A b C, and so on); only one kind could be formed by the recessive parent (a b c). In this case, eight kinds of phenotypically distinguishable progeny would be expected from such a test cross, and the numbers in each class should be approximately equal. For example, $\dfrac{A\ B\ C}{a\ b\ c}$, normal phenotype; $\dfrac{A\ B\ c}{a\ b\ c}$, normal wings and eyes, abnormal body pigment; $\dfrac{A\ b\ C}{a\ b\ c}$, normal wings, abnormal eyes, normal body pigment; and so on.

However, if the genes remain together on the same chromosome during meiosis, we would expect only two kinds of gametes from the hybrids, A B C and a b c; and only two kinds of progeny could be obtained when these gametes combined with the gametes formed by the recessive parent: $\dfrac{A\ B\ C}{a\ b\ c}$ and $\dfrac{a\ b\ c}{a\ b\ c}$. These two examples represent the extreme cases: segregation when the genes are on different chromosomes and linkage when they are on the same chromosome.

In an actual test cross, however, complete linkage is almost never found, because during meiosis a phenomenon called crossing over usually occurs. *Crossing over* is the exchange of genetic material between homologous chromosomes. It probably occurs when two homologous chromosomes break at the same place and reunite by exchanging equivalent sections. Crossing over seems to occur randomly along the length of the chromosome, resulting in the rearrangement of chromosomal genes into combinations different from those of the parent. Genes located far apart on a chromosome are more likely to be involved in exchanges than those located closely together, simply because the probability of breaks occurring between distantly separated genes is greater than the probability of breaks occurring between closely linked genes. In breeding experiments crossing over can be recognized by the appearance of progeny which have inherited a combination of linked genes not present in either parent. These offspring are called *recombinants*, and from the frequency of their occurrence we can estimate the frequency of crossing over between linked genes (see Figure 4.5).

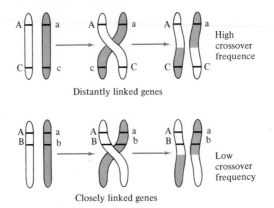

FIGURE 4.5 Cross over frequency as a function of gene distance. If two genes are relatively far apart, crossing over between them occurs frequently (top). If genes are close together, crossing over is much rarer. In general, the farther apart two genes are on the chromosome, the more frequent crossing over will be.

These consequences of crossing over were quickly realized by Morgan and his co-workers. By determining the frequency of crossing over between linked genes, they were able to determine the relative distances between them. By analyzing many pairs of linked genes in this manner, they were able to construct genetic maps of the chromosome. These genetic maps revealed that genes are located on the chromosomes in a linear order somewhat like beads on a string. More intensive genetic analyses have revealed that the number of groups of linked genes is always the same as the number of different chromosomes possessed by a given species.

Cytological examination of the chromosomes of various animals, as well as breeding experiments with *Drosophila*, also contributed directly to our understanding of how sex is determined. It was found, for instance, that female *Drosophila* have a pair of homologous sex chromosomes called the *X* chromosomes in addition to the three other chromosomes which are called *autosomes*. Normally male *Drosophila* have only one chromosome, but possess a differently shaped chromosome called *Y* with which it is associated. The fact that males and females are formed in approximately equal numbers can then be explained as follows. When the haploid eggs are formed by females, each egg receives one *X* chromosome; however, the male's haploid sperm cells contain either the *X* chromosome inherited from his mother or the *Y* chromosome inherited from his father. Those eggs fertilized by sperm-bearing *X* chromosomes develop into females (*XX*), and those bearing *Y* chromosomes develop into males (*XY*). In humans it appears that the *Y* chromosome itself is male determining, but in other species this is not necessarily so.

Human males are usually always *XY* and females are *XX*. However, failure of proper balancing of the sex chromosomes is well known, and individuals in which this occurs are often abnormal. For example, *XXY* individuals are underdeveloped males exhibiting Klinefelter's syndrome; individuals possessing an *X* chromosome but lacking an additional *X* or *Y* chromosome (*XO*) are underdeveloped females demonstrating Turner's syndrome.

4·3 Mutation and bacterial inheritance

The experimental findings that we have discussed demonstrated that characteristics inherited by the offspring from their parents are determined by units of inheritance called *genes*, which are located on or are an integral part of the chromosomes. Furthermore, it was shown that the chromosomes are not immutable structures, since rearrangements between homologous chromosomal pairs are a frequent occurrence. These studies emphasize the orderly mechanism for the transmission of *unchanged* hereditary characteristics. This is, of course, very important, for it takes thousands of years of evolution to accumulate the biological experiences of the race, and we would not want to lose this in a moment just because the information carrier is unreliable. It is also true, however, that if there were no changes in heredity, evolution would be impossible because the raw material for evolution is new or altered genes, and the only known process for obtaining them is mutation. *Mutations* are simply defined as "sudden heritable changes," and *mutagenic agents* or mutagens are anything that provokes their appearance.

Before discussing mutations and how they are brought about it is first necessary to define genes more adequately. Whereas it was quite certain that genes are located on chromosomes it was, and still is, not possible to see them nor to visually identify a given gene. Instead it is necessary to identify a gene by what it does, that is, by its effects on the phenotype. However, most phenotypic effects, especially in higher plants and animals, are generally quite complex and are often influenced by any of a large number of genes. For instance, the way in which the round, full phenotype of Mendel's peas is controlled by a gene is not immediately apparent.

The problem of gaining detailed information about genes from observing their phenotypic effects is further complicated by a fact we have alluded to but have not implicitly considered. This is that inheritance of genes is not tantamount to inheriting a trait or characteristic. In hybrids possessing dominant and recessive genes, as we have seen, only the phenotype of the dominant gene is expressed. In other cases, genes are expressed only when certain conditions of nutrition, temperature, and so on, are met, regardless of their states of dominance or recessiveness. For example, most Caucasians have genes for the formation of brown pigment in the skin cells, and hence can develop suntans. Obviously, however, few coal miners would ordinarily display this phenotype even though they have the genetic potential to do so.

Therefore, although enormous contributions were made to our under-standing of inheritance by studies on complex phenotypes, a great number of questions could not readily be answered by these techniques. It soon became apparent that simpler forms of life were more likely to yield detailed informa-tion about how genes replicate themselves and control hereditary traits than would continued studies focused exclusively on higher plants and animals.

Unlike plants and animals, bacteria are normally haploid and reproduce by simple binary fission. In this form of asexual reproduction, new individuals are formed from old ones when the parent bacterium, after growing to a certain size, divides to form two new unicellular organisms. Normally both of the new bacteria should have exactly the same properties as the cell from which they were derived. It is also evident that problems of dominance and recessiveness do not arise in haploid bacteria and that any allele present can be immediately expressed.

Bacteria continue to grow and divide into two new individuals as long as their food supply and other conditions are satisfactory, often reaching a population density of a billion or more cells per cubic centimeter. Therefore it is a simple matter to obtain very large bacterial populations, all of which are descended directly from a single cell. These populations are called *clones*, and since each bacterium can grow and divide into two new bacteria in a time as short as 20 minutes, a clone representing more than 15 generations can be obtained in one working day. To observe an equal number of human genera-tions would require almost 300 years!

For these reasons bacteria early became favorite objects for studying mutations and mutagenic agents. As said earlier, the bacteria in a clone are usually identical since they inherit those characteristics present in the original cell. Occasionally, however, "mistakes" occur and bacterial mutants with properties different from those of their parents arise. These mutants continue to breed true and their changed potentials are expressed almost immediately since there are no dominant alleles to mask the mutated gene.

Bacterial mutants resistant to agents that kill the parental strains have been studied intensively. Many kinds of bacteria are commonly killed by low concentrations of various antibiotics such as streptomycin. However, if large populations of sensitive cells are exposed to this drug, a few resistant mutants are almost always found. Clones derived from these mutants are also resistant to its effect. By knowing the total number of bacteria present and comparing this number to the number of resistant bacteria capable of growth in strepto-mycin, the rate at which the streptomycin-resistant mutants arise can be esti-mated. Under normal conditions of growth, this mutation occurs only about once every billion cell divisions.

Studies conducted in the 1940s demonstrated that these mutations occur spontaneously, probably as a consequence of random "mistakes" during replication of the genetic material which each daughter cell receives from its parent. It was conclusively demonstrated that antibiotics do not act as mutagenic agents or otherwise induce or direct mutations to antibiotic resistance. Adding the drug to the bacterial population simply allows the detection of the small number of mutants present by killing the sensitive cells. It was also shown that although streptomycin-resistant mutants generally breed true, they can also mutate back to the original nonresistant state at about the same rate at which mutation to resistance occurs. This phenomenon, called reversion, is a general one and occurs in most mutants studied.

If mutation to streptomycin resistance occurs as a random mistake in the replication of the bacterial genetic apparatus, it might be expected that other mistakes would occur during the many separate acts of duplication required to produce a population of a billion bacteria. This prediction has been amply verified. Mutations affecting almost every detectable trait occur at frequencies at least as high as the one described. Some examples are shown in Table 4–1.

TABLE 4.1

EXAMPLES OF SPONTANEOUS MUTATION RATES

SPECIES	GENE	MUTATION RATE
E. coli (bacteria)	Streptomycin resistance	10^{-9}
	Penicillin resistance	10^{-7}
	Inability to utilize the sugar, lactose	10^{-6}—10^{-7}
Z. mays (corn)	Shrunken seeds	10^{-6}
	Purple color	10^{-5}
D. melanogaster (fruit fly)	White eye	4×10^{-5}
Man	Albinism	2.8×10^{-5}
	Hemophilia	3.2×10^{-5}
	Total color blindness	2.0×10^{-5}

We now know that bacterial genes are composed of DNA, which is a long polymer made up of thousands of repeating chemical units called *nucleotides*. Genetic information is encoded in this polymer according to the sequence in which the chemical units appear, somewhat similar to the way that words can be spelled out in Morse code by a series of dots and dashes. As long as the original chemical order is maintained, the newly synthesized DNA each daughter cell receives will have the same genetic information. If mistakes are made, such as inserting a nucleotide out of order, these mistakes are transmitted to the offspring and usually cause a different phenotype.

It is not yet completely clear how spontaneous mutations occur during gene replication, but we know they occur rarely because most mutation rates are quite low. This means that living organisms have evolved a process whereby a remarkable degree of fidelity is achieved in the chemical synthesis of new genetic information from the pattern provided by the old.

A large number of agents and treatments drastically alter the rate at which mutations appear. As one would probably predict, many of these mutagenic agents exert their effects directly on the DNA of which the genetic messages are composed.

Ionizing radiations were the first agents used that proved capable of causing mutations. In 1927 H. J. Muller, then at the University of Texas, demonstrated the mutagenic activity of x rays. For his discoveries on the induced mutagenesis of *Drosophila* Muller received the Nobel Prize in 1946. Ionizing radiations are powerful, energetic forms of radiation that interact strongly with the atoms of the materials which they strike and generally cause profound alterations in their chemical structure. Ionizing radiations include those produced by x-ray machines, particle accelerators, and by the disintegration of radioactive elements such as radium, or radioactive forms of uranium and strontium. It is this last category that causes profound concern about contamination of the environment with the radioactive residues from nuclear explosions. The radiations these compounds produce have proved to be mutagenically active and there does not seem to be a threshold below which they are ineffective. Since some of the radioactive compounds are preferentially incorporated into growing tissue where they continue to emit radiations for many years, the danger from this source is clearly apparent.

The first proof that some chemical compounds cause mutations was obtained from a study on the effects of mustard gas (of World War I notoriety) on *Drosophila*. This compound is also a powerful carcinogen (cancer-producing agent) in animals. In general, the same agents which induce mutations also provoke cancers. These correlated activities include the radiations previously discussed. After the discovery of the mutagenic activity of war gases, a great many other compounds were shown to possess mutagenic activity and a list of them would fill many pages. Some of the materials in this category are caffeine, hydrogen peroxide, nitrous acid, and various base analogs. The base analogs include several compounds whose chemical structures closely resemble the nucleic acid bases found in DNA. Because these analogs are so closely similar to the normal constituents of DNA, the cell takes them up along with its foodstuff and uses them, instead of the normal bases, to make DNA. Their presence in the genetic material induces high mutation rates as a result of errors in replication.

Mutagenic agents are generally so strongly reactive that they kill a large proportion of the organisms exposed to them. However, in all the cases considered, the frequency of mutations among the exposed survivors is from 10 to 10,000 times higher than the normal, spontaneous frequency. It should be emphasized that mutagenic agents do not produce any special class of mutants not normally found spontaneously. Furthermore, they do not provoke the appearance of any particular class of mutants in preference to any other. We conclude that the main effects of mutagenic agents is to increase the rate at which mutations appear, although in some cases the particular part of the gene affected may vary from treatment to treatment.

In addition to the antibiotic-resistant mutants discussed, other kinds of mutants have also been subjected to intensive study. These include mutants of disease-producing bacteria which show either increased, decreased, or complete loss of ability to cause disease. Bacterial or viral mutants that have lost the ability to cause disease are called "attenuated" and are frequently used as vaccines to protect potential victims against the disease caused by the normal strain. Examples of attenuated mutants used in this manner include those derived from polio virus, rabies virus, and the bacteria which cause tuberculosis.

We have considered bacterial mutants in a general manner, but before leaving the subject, it might be instructive to discuss in some detail the actual methods used in the laboratory for obtaining nutritionally impaired mutants. The main problem concerned in this task is the fact that these mutations, like most other mutations, occur very infrequently, so that in any given population only one such mutant in several million normal bacteria is expected. It is obviously impractical to test individually several million cells or clones in order to identify and isolate one kind of mutant. Therefore various procedures have been developed to increase the proportion of mutants among the normal bacteria and to facilitate the tests used to identify them. In general, these include (1) increasing the number of mutants by exposing the starting populations to efficient mutagenic agents, (2) increasing the proportion of mutants by selectively killing the parental, nonmutant members of the population, and (3) simultaneously testing a great many clones with the least number of manipulations.

As an example let us consider the bacterium *Escherichia coli*. This is a useful organism for such purposes since it ordinarily grows and reproduces well on a diet consisting solely of water in which are dissolved a few inorganic salts and a simple sugar like glucose. From such a mixture, called a minimal medium, *E. coli* manufactures all the myriad organic constituents it requires. The minimal medium can be solidified by adding the inert jelling agent, *agar*, to it, as described in Section 1·7. Normal *E. coli* which grow on the minimal medium are called *prototrophs*, and nutritionally impaired mutants which cannot grow are called *auxotrophs*. Auxotrophs generally fail to grow on minimal medium because their genetic defect results in an impaired enzyme which is required to manufacture one of the organic substances they require. They usually grow quite well, however, if this substance is provided to them as a supplement of the minimal medium.

To isolate auxotrophic *E. coli* mutants the prototrophic population is exposed to a mutagenic agent such as x rays or nitrous acid and the survivors, which consist mainly of prototrophs plus the few auxotrophs that have been induced, are grown in a liquid minimal medium containing penicillin. This is done to increase the proportion of auxotrophs and is effective because penicillin kills only growing bacteria. Since only prototrophs can grow in minimal medium, most, but not all, of them will be killed. The penicillin is removed and the remaining bacteria, which now consist of perhaps five to ten auxotrophs per 100 prototrophs, are put onto solid medium containing all compounds required for growth. The individual clones which develop on this medium are tested for auxotrophy by transferring a portion of each clone to minimal agar. If the clone consists of auxotrophic mutants, growth will not occur.

This procedure was greatly facilitated by an ingenious technique called replica plating which was devised by Joshua and Esther Lederberg. Instead of individually transferring a portion of each clone from one medium to another, they pressed the surface of the agar on which the individual bacterial clones were growing onto a piece of sterile velveteen cloth. Large numbers of the individual cells in each clone were retained in the fibers of the velveteen, and then when petri plates of sterile minimal agar were pressed onto the velveteen "stamp pad" some bacteria from each clone were transferred to the new medium. By comparing the positions of the clones on the original plate to the positions of clones that developed on the second plate, it was an easy matter to discriminate between auxotrophs and prototrophs. The auxotrophs could then be isolated from the original plate and tested further. This procedure is illustrated diagramatically in Figure 4.6.

4·4 Sex in bacteria

Once bacterial mutants had become available, it was possible to test for the occurrence of sexual reproduction in bacteria. Such a phenomenon had occasionally been hinted at by bacteriologists who had observed under the microscope rare pairs of bacteria in suggestively close proximity, but it had never been convincingly demonstrated until 1946. In that year J. Lederberg, a graduate student at Yale University, and his mentor E. L. Tatum published their convincing studies on genetic exchange in *E. coli*.[2]

Their basic experiment consisted of mixing together two different auxotrophic mutants. For convenience the genotype of the mutant parental strains may be represented as $A^+B^+C^-D^-$ and $A^-B^-C^+D^+$ in which the plus sign denotes the ability of the organism to synthesize a given growth factor, and the minus sign represents dependence on an added source of the factor.

These two mutants were tested for the ability to genetically cross by first growing each culture separately in a "complete" medium to which the re-

[2] For their pioneer studies in bacterial genetics, Lederberg and Tatum received the Nobel Prize in Medicine and Physiology in 1952.

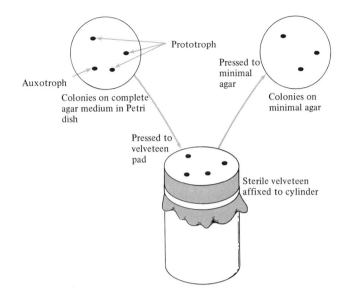

FIGURE 4.6 Replica plating. Colonies on complete agar plate are pressed onto the velveteen and some of the adhering bacteria are then transferred to a sterile minimal agar plate. After incubation the prototrophs form colonies and the auxotroph fails to grow. The auxotrophic colony can be isolated from the original plate for further study.

quired growth factors, C and D in the first case and A and B in the second case, were added. They were then washed free of the added growth factors mixed together, and the mixture was placed into a minimal agar medium free of any growth factors. In this medium only prototrophs could grow, that is, bacteria with the genotype $A^+B^+C^+D^+$, and clones of this type were found at the rate of about one for every million auxotrophic parents added to the mixture. The occurrence of prototrophs under these conditions could not be explained by mutation. Each parent required two different growth factors, and thus it would have taken two independent reverse mutations to produce an organism that could grow in the minimal medium. Since the frequency of reversion for each gene is approximately one in a million, the frequency of the double mutation, the product of the two events, is only one in 10^{12} bacteria, which is a million times less frequent than the rate that prototrophs actually appeared.

The occurrence of prototrophs could be explained, however, by assuming that the two auxotrophic mutants conjugated to form a fusion zygote, and that during reduction to the haploid state, portions of their chromosomes were exchanged by crossing over. This cross can be represented as

$$A^+B^+C^-D^- \times A^-B^-C^+D^+$$

and the fusion zygote as

$$A^+B^+C^-D^-$$

$$A^-B^-C^+D^+$$

It may be seen that a break between B and C on each chromosome followed by rejoining of the opposite halves will result in a recombinant chromosome containing the A^+B^+ markers from one mutant parent and the C^+D^+ markers from the other and that such a recombinant would be prototrophic for all of the genes under consideration.

After bacterial conjugation was discovered, it was studied in great detail by many scientists all over the world, with the result that today *E. coli* may well be Earth's best understood organism.

You may legitimately wonder why so many eminent scientists devoted their time and efforts to studying the sexual behavior of an insignificant creature such as *E. coli*. There are many reasons for their interest, including sheer, unqualified scientific curiosity—the desire to know. Furthermore, the ability to manipulate genetically any group of organisms has clearly great practical implications. Equally important, most biologists believe that all forms of existing life evolved from some simpler form of life. If this is so, we have no reason to think that evolution passed bacteria by nor that the forces of evolutionary adaptation have ceased to operate today.

It is obvious that evolution requires hereditary variation; that is, new varieties must arise which differ from those of previous generations. We stated earlier that mutation supplies, in the form of new or altered genes, the raw materials for evolution. Therefore we might assume that mutation alone could supply the variants required for evolutionary adaptation. However, many students of evolution doubt that the vast array of genes required for the coordinated activities of even the so-called simple bacteria could have evolved solely by random mutation. They point out that sexual reproduction provides an essential extra dimension of variability by allowing mixing or recombination of available (mutant and nonmutant) genes into a variety of genotypes. In this manner a particular gene can be mixed, for instance, by crossing over with a series of genes not previously associated with it. Those new gene combinations that are advantageous will be retained by natural selection, and deleterious combinations will be eliminated from the population. Therefore the demonstration that bacteria also sexually recombine had important implications which affect our understanding of evolution.

Once bacterial recombination was demonstrated, it also became of interest to determine the manner in which bacteria exchanged genetic informa-tion. As briefly described, higher plants and animals recombine sexually by forming specialized cells, the gametes, which help to ensure the orderly trans-mission of genetic information to the next generation.

Bacterial cells, however, differ profoundly from the highly differentiated cells of higher plants and animals. The cells of man, garden peas, and so on, are called *eucaryotic*, which means they have a true nucleus surrounded by a membrane and possess morphologically distinguishable chromosomes. Bac-terial cells are *procaryotic*, which signifies that they have no true nucleus. Their nuclear region, in which their DNA is concentrated, is not bounded by a membrane nor do they possess chromosomes. Instead bacterial genetic information is carried in one long piece of DNA, possibly 1000 times longer than the cells themselves. This long, narrow thread of DNA, which corre-sponds to the chromosomes of eucaryotic cells, is tightly compressed to form their nuclear region. It is still not completely understood how, in the absence of mitosis and meiosis, bacteria assure that each daughter cell receives an equal share of genes during asexual division, nor do we know in detail how genes are transmitted from cell to cell during sexual conjugation. However, the investigations, of which a summary follows, have revealed many of the details of sexual exchange in *E. coli*. It is hoped that a more complete under-standing of the genetics of the "simple" bacteria will help us gain new under-standing of genetic mechanisms of higher forms.

Escherichia coli is sexually differentiated into male donor cells, called F^+, and female recipients called F^-. Male bacteria differ from F^- recipients by possessing, in addition to their normal chromosomal DNA, a small circular piece of DNA called the F(ertility) factor. During conjugation only the F factor is efficiently transferred to the female, after which it maintains itself in the recipient by dividing each time the bacterial chromosome divides. Fe-males receiving F are thus converted to F^+ males and have all the properties of donor cells. The exact functions of the sex factor in cells that contain it are not known, but it seems to be involved in the formation of the conjugation bridge that connects mating pairs. The F factor can be eliminated from F^+ bacteria, and when this occurs, they behave in all respects as typical recipients (see Figure 4.7).

Normally F^+ males transfer only the F factor; however, mutants of F^+ cultures were found that efficiently transferred their chromosomal genes to the female recipients. These "supermales" were called Hfr for *high frequency of recombination*. When Hfr males were mixed with female recipients, proto-trophic recombinants were found at a rate 20,000 times higher than that re-ported in the original experiments of Lederberg and Tatum. The availability of these Hfr mutants made possible more detailed studies on the mating process in *E. coli* than would otherwise have been possible.

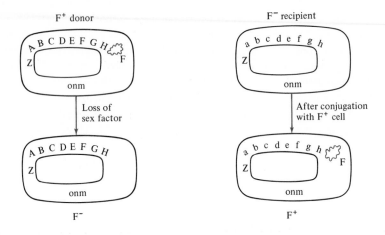

FIGURE 4.7 Two sexually differentiated bacterial cells. The bacterial chromosome is represented by the unbroken lines and genes are designated by letters. The wavy line represented the sex factor. During conjugation between F+ and F− bacteria, only the sex factor is transferred to the recipient, which is thereby converted to the donor state. The sex factor can be eliminated from F+ cells by various treatments and when this occurs they behave in all respects like typical recipients.

Many of the important experiments on the mating process in *E. coli* were made by E. Wollman and F. Jacob in Paris and W. Hayes in London. These workers mixed together Hfr males and F− females and at closely spaced intervals after mixing, they removed samples and interrupted the mating by subjecting the samples to high speed mixing in a Waring blender. In this manner they were able to determine which genes had been transferred to the F− by the Hfr at successive intervals after the two kinds of bacteria were mixed together. They found that soon after mixing, only one or two genes, for instance, A and B, had been transferred by the Hfr to the F−. When samples of the mating pairs were separated at slightly later times, one or two more genes, C and D, were found to have been transferred, and so on. These experiments revealed that a given Hfr mutant always transferred its chromosomal genes to the F− recipient in a definite, linear sequence. Usually, an Hfr mutant did not transfer its entire chromosome to the recipient, even in experiments where great care was taken to avoid separating the mating pairs. This was assumed to result because the conjugation bridge through which the Hfr chromosome is transferred is very fragile, and therefore mating bacteria generally spontaneously separate before chromosomal transfer is completed. As a result of this incomplete transfer of the male chromosome, the female recipients are only partially diploid. Crossing over can then occur between the entire female chromosome and that part of the male chromosome which was transferred before conjugation was interrupted.

Jacob and Wollman later isolated from a particular F⁺ bacterial strain, many different Hfr's. These were analyzed by the "interrupted mating" technique to determine the order in which they transferred their genes to the F⁻ recipient. Jacob and Wollman made the then surprising discovery that although each of these new Hfr mutants transferred their genes in a unique, oriented, linear sequence, the order of transfer varied from mutant to mutant. Thus Hfr 1 might transfer gene A first and Z last. This may be visualized as follows, with the arrow representing the direction of transfer: $\dfrac{A\ B\ C\ D\ E \ldots Z.}{}$

Hfr2, however, transferred its genes in the order $\dfrac{D\ E\ F\ G \ldots Z\ C\ B\ A,}{}$ whereas Hfr 3 transferred genes in the order $\dfrac{Z\ A\ B\ C\ D \ldots X\ Y,}{}$ and Hfr 4 had the order $\dfrac{Z\ Y\ X\ W\ V \ldots D\ B\ C\ A,}{}$ and so on.

It was also discovered that F⁻ cells are only rarely converted to the donor state after mating with Hfr males, and when the conversion to the donor state does occur, the converted recipients are always Hfr. This lack of ability to transfer efficiently the sex factor was shown not to be due to the loss of F by the Hfr's since they readily reverted to the F⁺ state and could again freely transmit the F factor to females. However, the reverted F⁺ cells could then transfer only the sex factor and could no longer transfer chromosomal genes.

These experimental observations were brilliantly explained by Jacob and Wollman in the following way. They pointed out that the different orders of gene transfer displayed by their Hfr mutants, which were all derived from the same F⁺ strain, could be explained best if it were assumed that the F⁺ chromosome was normally a circular structure. They hypothesized that the different Hfr mutants were formed when the unattached sex factor of the parental F⁺ strain became integrated into and was replicated as an integral part of the bacterial chromosome. According to their hypothesis, when the resulting Hfr mutant conjugated with the recipient cell, the male chromosome was broken at the point of sex factor attachment. The end of the chromosome opposite the attached sex factor was then the first to be transferred to the recipient. The different orders of transfer by their various Hfr mutants were explained to result because the sex factor became inserted at different sites on the bacterial chromosome during their formation from the F⁺ parent.

They explained that Hfr's are usually unable to convert F⁻ recipients to the donor state because the attached sex factor is the last part of the Hfr chromosome to be transferred during mating. Since chromosomal transfer is usually incomplete, most F⁻ recipients do not receive that part of the Hfr chromosome with the attached sex factor.

Several of these interpretations, which are now widely accepted, are presented in Figure 4.8.

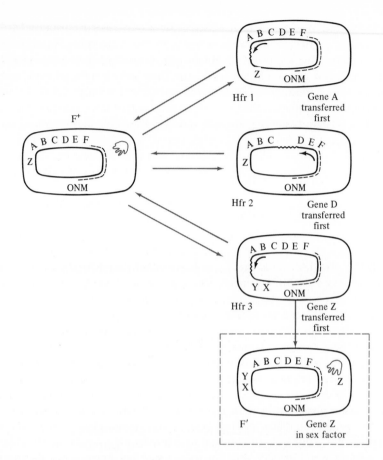

FIGURE 4.8 F⁺ bacteria give rise to Hfr cells when the sex factor becomes integrated into the bacterial chromosome. Integration can occur at any of many sites on the chromosome. When Hfr cells conjugate with F⁻ recipients, the order of chromosomal transfer by the Hfr is determined by the site of sex factor integration. Occasionally when the sex factor becomes released from the chromosomes it picks up some of the genes from the bacterial chromosome. The resulting F′ donors transmit their sex factor to recipients very efficiently, including the bacterial genes included in it.

Reversion of Hfr's to the F⁺ state is assumed to occur by reversal of the insertion process. When this occurs, the sex factor seems to once again replicate independently of the chromosome and is again freely transmissible to the recipients. Occasionally, however, when the sex factor is released from the bacterial chromosome, it incorporates adjacent bacterial genes into its structure. When these strains, called F′ (F prime) are mated with recipient F⁻ cultures, they transfer efficiently their sex factor to them. The recipients thus acquire the sex factor and also whichever bacterial genes were incorporated into it. This process whereby bacterial genes are transmitted from donor to recipient along with the sex factor is called *sexduction*. It was extensively employed to investigate the properties and control of enzymes involved in the utilization of the sugar, lactose, by *E. coli*. These studies are discussed in Chapter 8.

Genetic elements that like the sex factor can replicate either autonomously within the cell or become integrated into the bacterial chromosomes are called *episomes*. As we shall see, some bacterial viruses may also be classified as episomes, since they can also exist either connected to the bacterial chromosome or in the free state.

Additionally, other episomelike elements have been identified in bacteria. One example of these is called R factors (resistance factors). They were first identified in bacteria isolated during an epidemic of a disease characterized by severe intestinal disorders. Many bacteria isolated from patients with this disease were highly resistant to each of four different antibiotics to which these bacteria are normally susceptible. These resistant bacteria were able to transfer efficiently their resistance to susceptible strains. It was shown that antibiotic resistance is carried on a genetic element similar to the F factor and that it can be efficiently transmitted in a similar manner. However, the mechanism whereby resistance is gained by such a process is not clear.

In addition to conjugation, bacteria also exchange genetic information by two additional processes called *transformation* and *transduction*. *Bacterial transformation* is the process whereby purified DNA extracted from one kind of bacterium is taken up from the medium and is genetically incorporated into the chromosomes of recipient bacteria. It is not certain that transformation is an important means of genetic exchange in nature, but it has been an important tool for studies on the physical and biological properties of DNA. A detailed account of the discovery of transformation and studies of its nature is presented in Chapter 5. *Transduction* is a form of genetic exchange between related bacteria in which bacterial viruses serve as agents of transmission. We shall defer our discussion of transduction until after presenting a general description of viruses.

4·5 Viruses: Genetic parasites

In 1892 D. Iwanowsky reported the transmission of tobacco mosaic disease to healthy plants by juice taken from infected plants and filtered through bacteria-tight filters. Soon afterwards other diseases were shown to be transmissible by similar bacteria-free extracts, and the agents responsible were called *viruses*.

Diseases produced by viruses were recognized long before their biological nature was understood. Viruses are usually not large enough to be seen in the light microscope, and thus escaped detection until special procedures were devised for this purpose. Especially significant was the introduction of the electron microscope into biological research in the 1940s. In Chapter 2 a number of electron micrographs of viruses are presented, and the student is urged to review frequently these pictures as he studies this and the following sections.

Viral diseases of man include poliomyelitis, influenza, the common cold, measles, mumps, smallpox, and rabies. Viruses also cause serious diseases in domestic and wild animals, such as hoof-and-mouth disease in cattle and horses, distemper in cats, and various diseases of fowls. Almost all plants are subject to viral diseases and symptoms induced vary from lesions on the leaves to severe stunting of growth. Despite intensive investigations on plant and animal viruses, knowledge concerning their essential biological properties was difficult to obtain. This was because, in addition to their small size, viruses cannot multiply outside the cells of their host. Therefore investigators studying viruses were forced to use whole plants and animals for growing them, with the attendant difficulties involved in obtaining and caring for the plants or animals required in their experiments. Despite these inherent difficulties, the first advances in our knowledge of viruses were made with these materials. It was shown that most viruses are relatively *host-specific*, that is, a given virus normally infects only one or a few closely related species. Wendell Stanley, now Director of the Virus Laboratory, University of California, Berkeley, was able to purify and concentrate tobacco mosaic viruses until he obtained aggregates of them in crystalline form. This allowed careful chemical analyses on these viruses which showed that they contained about 95 percent protein and 5 percent ribonucleic acid, and nothing else.

However, detailed studies on the multiplication of viruses within the cells of the host they infect were not feasible in the complex environment of whole plants or animals, and consequently much of what we know about viruses was first learned from studies on bacterial viruses (bacteriophage). This was because the effects of bacteriophage on their unicellular bacterial hosts were technically much easier to determine than were the interactions of other viruses with their more complex hosts.

4·6   The life history of bacteriophages: Virulent and temperate

Bacteriophage (or phage for short) were discovered about the time of World War I and were recognized by their ability to dissolve, or lyse, bacterial cultures. It was first thought that these agents would be important for the prevention and treatment of bacterial diseases. A fictional but relatively accurate account of the discovery of bacteriophage and of efforts to use them to treat a plague epidemic is included in *Arrowsmith*, a novel by Sinclair Lewis. As in his story, phage therapy did not prove effective, but studies on the biological properties of bacteriophage contributed greatly to our understanding of the chemical and biological interactions of viruses and living cells.

One of the most dramatic demonstrations of the effects of bacteriophage on their bacterial hosts occurs when a small amount of bacteriophage suspension is added to a liquid medium, visibly teeming with bacterial growth. Within an hour, under proper conditions, the turbid bacterial culture starts to bubble and froth, and within a few minutes no visible trace of bacterial life remains. The bacteria have been dissolved, or lysed, and the cleared culture liquid can now be shown to contain manyfold more bacteriophage than were originally introduced into it. In some manner the bacterial viruses have multiplied at the expense of their hosts.

The number of phage particles or infected bacterial cells in a suspension is easily determined by the plaque method. This consists of appropriately diluting the suspension and adding a small amount of it to a tube containing a few drops of a concentrated bacterial culture. A small amount of warm, molten nutrient agar is then added, and the entire mixture is poured onto the surface of a petri dish containing a nutrient agar medium. Each infected cell is surrounded by a large number of uninfected cells which soon grow into a continuous sheet of cells called a *lawn*. During this time the infected cells produce progeny phages which are released and start new cycles of infection in the surrounding cells. After several such cycles of infection, a hole appears in the lawn whose position corresponds to that of the original infected cell. These holes, called *plaques*, are counted and their number is related directly to the number of phages in the original suspension.

The sequence of events between the time bacteriophage are added to a bacterial culture and new bacteriophage appear is now fairly well known. This body of knowledge is the result of the efforts of a great many workers during the past 40 years. During the course of obtaining this information, important biological principles were established, as well as an understanding of the niche which viruses occupy in the biological hierarchy.

The modern era of bacteriophage research was started by Max Delbruck at Cal Tech about 1940 and was soon being carried on by many other researchers in the United States and elsewhere. Much of this research was confined to studies on a group of seven phages which infect *E. coli* and are named T_1 (for Type 1), T_2, T_3, and so on. By coincidence, the even-numbered T phages (T_2, T_4, and T_6) happened to be closely related and quite similar to one another. These T-even phages have been most intensively studied and will serve as examples in our discussion of the virulent phage. *Virulent* phages are those which invariably lyse the cells they infect during the process of producing progeny phages.

The T-even phages are chemically and morphologically indistinguishable. They are composed almost exclusively of protein and DNA, which are present in approximately equal amounts. They have a hexagonally shaped head to which a slender tail is attached, giving them somewhat the appearance of tad-poles. The DNA of these phages, which corresponds to their chromosome, is one continuous strand about 50 microns long (this is about 25 times longer than an *E. coli* cell, but is only about $\frac{1}{30}$ as long as the bacterial chromosome). The phage DNA is tightly condensed in a protein-covered head. The tail is encased in a contractile sheath and has six tail fibers attached to its base.

Other phages have been discovered that are somewhat larger than the *T* phages, and a great many have been found that are smaller. Many of the smaller phages have no tail structure and are essentially spherical in shape; among these some have been shown to contain RNA as their genetic material instead of DNA.

The first step in the infection of a sensitive bacterium with a virulent phage occurs when the nonmotile phage collides with a bacterium as a result of random motion. The phage adsorbs to the cell by its tail fibers which inter-act with specific chemical sites on the bacterial surface. When this occurs, the phage is irreversibly adsorbed to the cell. Next a hole is made in the cell wall, probably by an enzyme in the phage tail. Then the tail core, a hollow needlelike tube which extends the length of the tail, is introduced into the hole by contraction of the sheath. The DNA contained in the phage head is then injected through the core into the bacterium. After the DNA has penetrated into the cell, the proteineacous phage components, which remain outside the cell wall, can be sheared off without affecting the outcome of the infectious process. Experiments which proved that only the nucleic acid enters the bacterium were performed by Alfred D. Hershey and Martha Chase and are presented in detail in Chapter 5.

After the phage DNA has been injected into the susceptible bacterium, it starts to multiply by a process called *vegetative growth*. Vegetatively multi-plying phage DNA is not infectious, even if infected cells are broken open artificially and their contents assayed. Soon after infection, synthesis of bac-terial products stops and the bacterial biosynthetic machinery is diverted exclusively to the manufacture of new phages. The phage's genetic information is substituted for that of its host cell (which is destroyed by virulent phage) and contains instructions for the synthesis of new phage DNA and protein components and for their assembly into infectious particles.

The sequence of events after infection is as follows. In about 5 minutes new enzymes are made by the cell's synthetic machinery according to specifica-tions from phage genes. These enzymes catalyze the synthesis of new copies of phage DNA. This is followed shortly afterward by the synthesis of phage structural proteins such as those making up the head, tail, core, sheath, and fibers. In about 15 minutes the completed phage chromosomes are condensed and encapsulated within a new protein covering to make the new phage heads. The newly synthesized tail components are then added to the preformed heads, thus completing the new infective particles. When about 100 to 200 new phages are assembled in the infected bacterium, a phage-directed enzyme dissolves the cell wall and releases the mature progeny phage which are now able to initiate new infectious cycles.

In addition to virulent phages such as the *T*-evens, other less virulent kinds called *temperate bacteriophages* have been studied. When temperate phages infect a susceptible bacterium, a certain proportion of the infected cells lyse and new phages is released in a manner similar to virulent phage infection. A certain proportion of infecting temperate phages, however, become reduced to a latent state (called *prophage*) and instead of destroying the cell, they are maintained within the surviving bacteria. Bacteria containing prophage are called *lysogenic*.

Lysogenic bacteria differ from nonlysogenic bacteria in only a few properties. They are immune to further infection by the same kind of phage they carry, and they have the hereditary potential for phage production in the absence of external infection. When a culture of lysogenic bacteria is treated with various agents such as nitrogen mustard, ionizing radiations, or ultraviolet light, they are induced. As a result of *induction* the prophage is released from the lysogenic bacterium's chromosome and multiplies like a virulent phage. Each induced cell produces normal phage progeny, and the mature particles released from them have the same properties as any other temperate phage.

Lysogenic bacteria are formed when the temperate phage DNA, after injection into the cell, becomes inserted into the bacterial chromosome at a particular location. After insertion into the bacterial chromosome, the prophage is inactive except for a limited number of its genes. It is replicated along with the bacterial chromosome and is distributed with it to each daughter cell prior to cell division (see Figure 4.9).

4•7 Transduction

After the initial discovery of genetic exchange in *E. coli*, sexual recombination was demonstrated in a number of other bacterial species. One of the species exhibiting genetic recombination, *Salmonella typhimurium*, was studied in detail by N. D. Zinder and J. Lederberg. (This time Lederberg was the mentor.) As with *E. coli*, they found that by mixing together two different mutants, offspring were obtained which contained genes from both parents. To test whether direct contact between parents was necessary, they applied the "U-tube" test. Before relating the results of their experiment, let us briefly discuss the U-tube test and its application to *E. coli*.

The U-tube (Figure 4.10) is a device consisting of two arms separated by a sintered-glass filter whose pore size is just small enough to hold back bacteria. This device was first utilized by B. Davis in 1950 to test whether true sexual fusion is necessary for genetic exchange in *E. coli*. After placing the

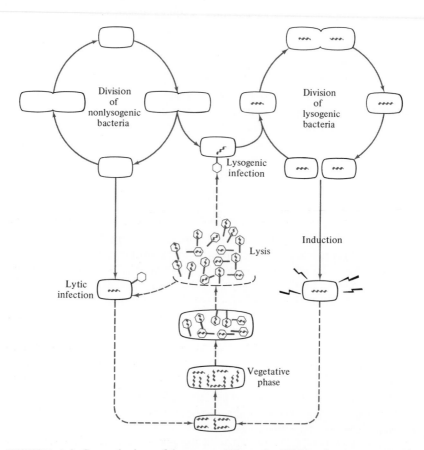

FIGURE 4.9 General view of lysogeny. When the DNA of a temperate virus enters the cell, it may either immediately multiply like a virulent virus or be changed into the prophage. Each descendent of a lysogenic cell is also lysogenic. Induction with radiations or chemicals causes the prophage to multiply and results in the lysis of the cell following production of new temperate phages.

parental strains of *E. coli* in the different arms of the tube, the medium was flushed through the filter from one side of the tube to the other. Since no recombinant offspring were obtained under these conditions, Davis concluded that cell-to-cell contact was necessary for genetic exchange in *E. coli*.

When this test was applied to *Salmonella* by Zinder and Lederberg, however, recombinants *were* formed. But only in one arm of the tube. This suggested to them that some product from the donor strain was passing through the glass filter and transforming the recipient parent to a new genotype. After a series of clever experiments, Zinder and Lederberg discovered that the filterable product was a temperate phage.

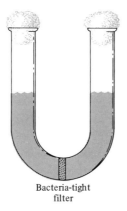

Bacteria-tight
filter

FIGURE 4.10 The U-tube is used to test if cell-to-cell contact is necessary for sexual recombination in bacteria.

The process whereby temperate phage transfer bacterial DNA from one bacterium to another is called *transduction*, and the phages which carry the bacterial genes are called *transducing phages*. Transducing phages are found at a frequency of about one per million normal phages. They are formed when the temperate phages infect the bacterial cells, enter into the phase of vegetative growth, and start encapsulating their newly synthesized DNA into phage particles. Occasionally, a piece of the bacterial DNA, which in these cases is broken down into small pieces, is encapsulated into the head of an otherwise normal phage particle. The adventitious inclusion of bacterial DNA in phage particles occurs very infrequently, but seems to occur equally well for any section of the bacterial chromosome. As a result, when recipient bacteria are infected with these transducing particles, bacterial DNA is injected into them. Different particles contain different sections of the bacterial chromosome, any of which can be integrated into the recipient's chromosome by a crossover event.

Note that the net results of transduction and conjugation are very similar. In both cases a piece of bacterial DNA from a donor is inserted into a recipient which possesses its own complete chromosome. A section of the recipient's chromosome can then be replaced by crossing over with the corresponding section of the inserted donor DNA. The main differences in these two forms of genetic exchange, other than the obvious differences in the means of DNA delivery, are that the transducing DNA fragment is smaller than the DNA injected by an Hfr and sexually differentiated bacteria are not required for transduction. Some of the details of transduction are presented in Figure 4.11.

An investigation of the factors which contribute to the ability of *Corynebacteria* to cause diphtheria led to the discovery of still another way in which

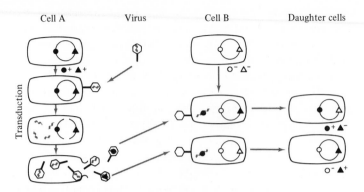

FIGURE 4.11 In transduction the agent for transferring bacterial genes is a virus particle. The bacteriophage injects its genes (wavy line) into bacterial cell A, and the genes create new copies of the bacteriophage. Occasionally the new bacteriophage particles so formed enclose a few genes from the chromosome of the bacterial host along with a few viral genes. These imperfect bacteriophage are able to inject their contents into another cell (cell B) but are unable to destroy it. In this way genes (solid black shapes) can be transferred from cell A to daughters of cell B.

phage can alter the hereditary properties of bacteria. The clinical symptoms of diphtheria result from a toxic substance produced by *Corynebacteria* when they grow in the throat of an infected person. The toxin which these bacteria produce is carried to other parts of the body and acts as a powerful cell poison. Recently it was shown that all toxin-producing strains of *Corynebacteria* are lysogenic for a particular prophage. *Corynebacteria* which were not lysogenic for this phage did not produce toxin nor did they cause diphtheria, even though they could grow in the throat. These nonlysogenic *Corynebacteria* could be converted to toxin-producers by infecting them with the phage, and, of course, they were now capable of causing diphtheria. This process, whereby nontoxigenic *Corynebacteria* become toxin-producers as a result of phage infection, is called *conversion*. In conversion the factor which confers toxigenicity to lysogenic bacteria seems to be a normal part of the phage's genetic makeup, and not a bacterial gene which is picked up by a few phage particles as in transduction.

4·8 Viruses also have genetics

Phages not only carry genes from one bacterium to another but they also have an efficient mechanism of genetic recombination themselves. When two genetically distinguishable phage mutants simultaneously infect the same bacterium, both kinds replicate within it and approximately equal numbers of both kinds of progeny are produced. During vegetative replication following mixed infection, both phage chromosomes, prior to being encapsulated in the protein head, come into intimate contact and can exchange segments by crossing over. To demonstrate genetic recombination in bacteriophage, two phage mutants can be used. One mutant forms large round plaques that are clear and the other mutant forms small turbid plaques. The genotypes of these mutant phages may be represented as r^-tu^+ and r^+tu^-, respectively. When these two mutants simultaneously infect the same cell, four kinds of progeny phage are produced: the parental phages r^-tu^+ and r^+tu^-; additionally recombinant types, r^-tu^- which produce round turbid plaques and r^+tu^+ which produce small clear plaques, are also found. Each of these four kinds continues to produce progeny with these characteristics in subsequent single infections; in other words, they breed true. Under standard conditions of mixed infection, the frequencies at which the four kinds of phage are produced are quite reproducible. Similar kinds of crosses between large numbers of different phage mutants have demonstrated that viral chromosomes can be mapped by procedures not greatly different from those used by Mendel and Morgan in their studies of inheritance in garden peas and *Drosophila*.

Thus we have seen that all cellular properties are determined by genes which are organized in a linear array called chromosomes. We have seen that the same general features govern inheritance in all biological systems, and are led to a belief that at the genetic level many common processes link us all together.

QUESTIONS AND PROBLEMS

4.1 Give the fundamental difference between

   (a) mitosis and meiosis
   (b) transduction and conjugation
   (c) mutation and selection
   (d) virulent and temperate phage
   (e) genotype and phenotype
   (f) bacteria and viruses
   (g) conjugation and mutation

4.2 Evolution depends on organic diversity. What is the physical basis of organic diversity? What are some agents that can induce diverse forms?

4.3 The F_1 generation of a cross between a short plant producing red flowers and a tall plant producing white flowers were all tall with red flowers. If members of the F_1 generation were crossed with each other, what phenotypes would be expected among the F_2 generation and in what proportion? (Assume the genes controlling these factors are on separate chromosomes.)

4.4 A large number of antibiotics have been found which can be used to control infectious diseases caused by bacteria. Why do you suppose none has been found which is useful in treating diseases caused by viruses?

4.5 What are the arguments, pro and con, for the proposition that *viruses are living*?

4.6 You have two strains of *E. coli*:

Type A: Resistant to streptomycin, sensitive to penicillin, and requires sugar for growth;

Type B: Resistant to penicillin, sensitive to streptomycin, and requires sugar for growth.

The following experimental results were obtained:

10^8A cells placed on sugar and penicillin yields 100 colonies.

10^8A cells placed on streptomycin with no sugar yields 10 colonies.

What is the mutant frequency of penicillin sensitive to penicillin resistant? How many colonies would you expect if you placed 10^9 B cells on penicillin with no sugar? If you placed 10^9 A cells on penicillin and no sugar?

4.7 Recently, Salvador E. Luria, one of the pioneers of bacteriophage research, defined viruses as "entities whose genome is an element of nucleic acid, either DNA or RNA, which reproduces inside living cells and uses their synthetic machinery to direct the synthesis of specialized particles, the virions, which contain the virual genome and transfer it to other cells." What do you think the most important parts of this definition are? What evidence can you cite for the various parts of this definition?

SUGGESTED READINGS

Fraenkel-Conrat, H., *Design and Function at the Threshold of Life: The Viruses*. New York: Academic Press, Inc., 1962. A paperback introduction to tobacco mosaic virus and the general question: Are viruses living?

Hayes, W., *The Genetics of Bacteria and Their Viruses*, 2d ed. New York: John Wiley & Sons, Inc. 1968. A well-written reference textbook which is used by professional bacterial geneticists.

Jacob, F., and E. L. Wollman, "Viruses and Genes," *Scientific American*, June 1961 (offprint 89). An illuminating discussion of both heredity and infection.

Peters, J. A. (ed), *Classic Papers in Genetics*. Englewood Cliffs, N.J.: Prentice-Hall, Inc., 1959. Especially pertinent are the translation of the original classic paper of Mendel, a short paper by Muller on induced mutation, and the original reports of sexual recombination in bacteria by Lederberg and Tatum and Zinder and Lederberg.

Sonneborn, T. M. (ed), *The Control of Human Heredity and Evolution*. New York: The Macmillan Company, 1965. This book contains a series of essays on the moral, ethical, and social problems that have been raised by recent discoveries in genetics.

Wills, C., "Genetic Load," *Scientific American*, March 1970. A stimulating discussion of mutation and the viability of the species.

Wollman, E. L., and F. Jacob, "Sexuality in Bacteria," *Scientific American*, July 1956 (offprint 50). A lucid discussion of conjugation by two of the men who have made great contributions to the field.

Wood, W. B., and R. S. Edgar, "Building a Bacterial Virus," *Scientific American*, July 1967 (offprint 1079). A recent account of the assembly of viral components in the test tube.

5

THE GENETIC MATERIAL: DEOXYRIBONUCLEIC ACID

Philosophy begins when someone asks a general question, and so does science.

BERTRAND RUSSELL

In Chapter 4 the concept of a gene was presented in order to explain certain patterns of inheritance. The gene was characterized as an information packet which is located at a particular point on a particular chromosome. In effect, the genes were treated as though they were strings of beads of unknown composition, which are duplicated and transmitted from parent to offspring by an unknown mechanism and which somehow interact with the environment to determine the phenotypic characteristics of the organism. We are now ready to ask the following three crucial questions:

1. What is the chemical nature of the gene?
2. How are the genes reproduced so that each offspring gets an (almost) identical copy?
3. How is the genetic information converted into the observable physical trait?

In this chapter we will study some beautiful experiments which provide the answers to the first two questions. The third question will be discussed in Chapter 6.

5·1 The discovery of transformation

In 1928 Fredrick Griffith, a medical officer in the British Ministry of Health, reported the results of his studies on what, at that time, was the number one killer of man, pneumonia. It was known then that pneumonia was caused by a spherically shaped bacterium, the pneumococcus. Griffith began his studies by isolating the bacteria from the sputum of patients with pneumonia. When the bacteria was first obtained in pure culture, he found there was only one type of pneumococcus, later called type *S*. This type was characterized by the appearance of mucoid or smooth (*S*) colonies on agar medium; when type *S* bacteria were injected into mice, the mice contracted pneumonia and died. When type *S* bacteria were subcultured by transferring from one agar plate to another, there occasionally appeared (we now know, by mutation) a second type of pneumococcus, type *R*. This type gave irregularly shaped or rough (*R*) colonies and was unable to cause the disease.

Griffith realized that the basic difference between the *R* and *S* strains is that type *S* contains a polysaccharide capsule surrounding the cell. These capsules help produce the smooth appearance of the colony, and more important, protect the bacteria so that they can survive and grow in the animal body. Type *R*, lacking the capsule, gives rise to a rough colony and cannot survive inside the animal.

To understand how these capsules participated in the infectious process, Griffith performed a series of experiments on mice. Keep in mind that the motivation for these experiments was not a curiosity regarding the nature of the gene but rather a quest for information about the nature of the infectious process. As so often happens in science, the correct interpretation for these experiments came much later. For the moment we shall consider only Griffith's experimental results; the interpretation and significance will emerge as subsequent experiments are discussed. Table 5–1 summarizes the results of Griffith's experiments.

TABLE 5.1

GRIFFITH'S ORIGINAL TRANSFORMATION EXPERIMENTS

EXPERIMENT	INJECTION INTO MICE	RESULT
1	Type *R* only	No effect
2	Type *S* only	All mice dead
3	Heat-killed type *S*	No effect
4	Type *R* plus heat-killed type *S*	All mice dead

Experiments 1 and 3 were controls. Neither type *R* nor heat-killed *S* was able to cause the disease when injected separately. However, when they were mixed together and then injected, the concoction was lethal. Furthermore, a *postmortem* on the mice from experiment 4 revealed *only live type* S *pneumococci*. These type *S* bacteria bred true; that is, the smooth characteristic was retained and reproduced in subsequent generations. The process by which type *R* bacteria were converted into type *S* came to be known as *transformation*. Although Griffith performed careful controls and his type transformation was confirmed by other workers, it did not attract due attention. One possible reason for the apparent lack of interest in Griffith's experiments was because he did not emphasize the genetic implications of his work. In 1942 Griffith was killed in London during a bombing raid and he did not live to see the significance of his experiments appreciated.

5·2 Transformation: The chemical nature of the gene

In the 1930s a group at the Rockefeller Institute led by Oswald T. Avery (1877–1955) became interested in the process of transformation. At first Avery was skeptical of Griffith's work and asked one of his young assistants to repeat the experiments. Not only were Griffith's experiments reproducible, but it was soon discovered that mice were not needed to demonstrate transformation. In these later experiments the bacteria were treated, mixed, and placed directly on the appropriate medium. The results were approximately as shown in Table 5–2.

TABLE 5.2

AVERY'S TRANSFORMATION EXPERIMENTS

EXPERIMENT	NUMBER AND TYPE OF BACTERIA TREATED	NUMBER AND TYPE OF COLONIES PRODUCED
1	10^6 (1,000,000) type R	1 type S, 10^6 type R
2	10^6 heat-killed type S	0
3	10^6 type R + 10^6 heat-killed type S	10^4 type S, 10^6 type R

Experiments 1 and 2 were controls. In experiment 1, only one out of a million of the R bacteria gave rise to type S colonies. The probable cause of this infrequent event is spontaneous mutation which was discussed in Chapter 4. Experiment 2 demonstrated that all of the bacteria were killed by the heat treatment. In experiment 3, one out of 100 of the bacteria were converted from type R to type S. Since this frequency is much higher than the mutational rate, a process of transformation must have taken place.

It is reasonable to ask at this time: Why, when the mixture of heat-killed S and live R was injected into a mouse, were *all* of the bacteria found at the *postmortem* type S, whereas when placed directly on the agar medium, only 1 percent were type S? The answer is that in all probability Griffith also obtained only 1 percent transformation, but that when injected into the mice, the type R were destroyed and did not show up in the postmortem.

Avery and his collaborators concluded from these experiments that heat-killed type S contain an active substance which enters into the live type R and transforms it into type S. Since this trait is inherited in subsequent generations, it has the property that we uniquely associate with the gene. This active substance was called *transforming principle*. The Rockefeller group then began an extensive program to isolate and identify the transforming principle.

Biochemists have developed, mostly by trial and error, a series of techniques for isolating and identifying chemical substances from cellular matter. The first step in an isolation procedure is to prepare a cell-free extract by disrupting the cells. This can be accomplished in a number of different ways, such as freezing and thawing, grinding with powdered glass, or by various chemical treatments. In the case of pneumococcus, the cells were broken by treatment with the chemical, sodium deoxycholate. Was the transforming principle still active? It was. When the cell-free extract from type *S* was added to live type *R*, approximately 1 percent of the cells were transformed to type *S*. The next step was the separation of the components which are released when the cell is disrupted. By a variety of techniques the chemicals can be separated according to their size, electrical charge, and other properties. In the case of transforming principle, the key step in the isolation process was the use of ethyl alcohol. When two volumes of 95 percent alcohol were added to one volume of the cell extract, a stringy white precipitate formed in the solution. This precipitate was easily collected on a glass rod, leaving most of the other components behind either in the alcoholic solution or as granular precipitates. *All of the transforming principle was associated with the stringy precipitate.* The procedure was repeated a number of times until Avery was satisfied that he had the transforming principle in as pure a state as he could get it.

The next question: What is it? The observation that the material precipitates in alcohol indicates that it is a large molecule, a macromolecule. But is this stringy white precipitate a protein, deoxyribonucleic acid (DNA), ribonucleic acid (RNA), or polysaccharide?

With a few precious millgrams of the highly purified transforming principle, the Rockefeller group began their analyses. First, they determined the elementary composition of their unknown substance. It contained 36 percent carbon, 4 percent hydrogen, 16 percent nitrogen, 10 percent phosphorus, and 34 percent oxygen. This elementary chemical analysis, especially the relatively high phosphorus content, suggested that the substance was a nucleic acid, either DNA or RNA. As discussed in Chapter 1, proteins and polysaccharides do not contain appreciable amounts of phosphorus. The two types of nucleic acid can be distinguished by the type of sugar they possess. All DNA molecules contain the sugar, deoxyribose, whereas RNA molecules contain the sugar, ribose. Since the sugar analyses revealed that the unknown substance contained only deoxyribose, the transforming principle was evidently DNA.

It was still possible, however, that a minor component not detected by the analyses was responsible for the transforming activity. Therefore as a final demonstration that the active substance was DNA, a series of enzyme studies was conducted. The three enzymes employed were trypsin, ribonuclease (RNase), and deoxyribonuclease (DNase). Trypsin is a specific enzyme in that it only destroys protein; likewise, RNase destroys only RNA, and DNase destroys only DNA. The transforming principle was treated with each enzyme and then tested for activity (Table 5–3).

TABLE 5.3

EFFECT OF SPECIFIC ENZYMES ON TRANSFORMATION

EXPERIMENT	TREATMENT OF TRANSFORMING PRINCIPLE	RESULT OF TRANSFORMATION EXPERIMENT
1	Trypsin	Transformation achieved
2	RNase	Transformation achieved
3	DNase	No transformation

Thus in 1944, after 15 years of persistent research, the Rockefeller group had clearly demonstrated that the transforming principle was DNA. Since the transforming principle has the property of a gene, and the transforming principle is DNA, *the gene must be made up of DNA* (at least in this case).

By a mechanism which we still do not understand, minute amounts of DNA enter the permissive cell and become permanently established. These recipient cells now act as if a new gene had been added to their genetic makeup. In Griffith's original experiments, the transformation process was made possible by the fact that the temperatures required to kill pneumococci (65°C) do not destroy the DNA. Thus DNA from the heat-killed type *S* entered the type *R* cells and transformed them into type *S*.

Avery's extraordinary results were soon enlarged on in two significant ways. First, it was shown that a large number of other traits could also be transformed. For example, DNA extracted from strains of pneumococci which were resistant to the drug streptomycin could transform streptomycin-sensitive pneumococci into the resistant form. Second, the process of transformation was demonstrated in two other species of bacteria, *Bacillus subtilis* and *Hemophilus influenzae*. Again it was shown that the active transforming principle was DNA. By 1950 the evidence from these and other experiments was convincing that in bacteria, at least, the genetic material is DNA.

5·3 DNA is the genetic material in viruses, animals, and plants

Although the transformation experiments clearly demonstrated that the gene is DNA in certain bacteria, there was at first some skepticism about the general applicability of Avery's discovery. Much of this skepticism vanished when it was shown by Alfred D. Hershey and Martha Chase of the Cold Spring Harbor Laboratories in New York that DNA was also the genetic material in viruses. These investigators used for their critical experiments a simple bacterial virus called phage T_2. In order to understand the Hershey-Chase experiment, you must remember (Section 4·7) that (1) phage T_2 is composed of only two kinds of macromolecules, protein and DNA; (2) pictures taken with an electron microscope (review Figures 2.32 to 2.34) reveal that only part of phage T_2 enters a bacterium; once inside, the phage multiplies until a few hundred *complete* new phages are produced. Thus that part of the phage which enters the bacteria must contain the genetic information for all of the phage parts. The question then is, what part of the phage enters the bacterium?

Hershey and Chase prepared for their experiment by growing one batch of bacteria on a nutrient medium containing radioactive phosphorus atoms (^{32}P); another batch was grown on nutrient medium containing radioactive sulfur atoms (^{35}S). Each batch was then infected with phage T_2. As the phage grew inside the bacteria, some of the radioactivity was incorporated into its structures. The ^{35}S went exclusively into the protein part of the phage since DNA contains no sulfur. The ^{32}P went exclusively into DNA since phage proteins contain no phosphorus. In this way the protein and DNA components of the phage were selectively labeled with radioactivity.

The ^{32}P and ^{35}S phages were then used to infect separate cultures of fresh bacteria. This time, however, there were no radioactive atoms in the medium. After allowing just enough time for the phages to infect the bacteria, the mixtures were shaken vigorously (they used a Waring blendor) and the bacteria purified. The shaking was necessary to dislodge the parts of the phage which became attached to the bacteria but which did not actually penetrate it. Thus only the part of the phage which entered the bacterium remained with it during purification. The purified bacteria were then examined for radioactivity. If radioactivity were found with the bacteria infected with the ^{32}P phages, it would mean that DNA penetrated the bacteria; if radioactivity were found with ^{35}S phage infected bacteria, it would mean that the proteins penetrated. The result was that only the bacteria which were infected with ^{32}P phages were radioactive. *Therefore the only part which enters the bacteria and must contain the genetic information is DNA.* As was discussed in Chapter 4, the phage protein serves only to aid the DNA in entering the cell and in protecting it against injury.

The Hershey-Chase experiment completed in 1952 was the first clear demonstration that DNA was the genetic material in a virus, phage T_2. Since that time experiments have shown that DNA is also the genetic material in a number of other viruses. William R. Romig has recently performed such an experiment at the University of California at Los Angeles. In this case the experiments were performed with a virus called SP8 which infects the bacterium *B. subtilis*. As mentioned previously, *B. subtilis* is one of the few kinds of bacteria that can carry out the process of transformation. Romig's technique was to mix purified DNA isolated from phage SP8 with its bacterial host, *B. subtilis*. The phage DNA entered the bacteria in the same manner as transforming principle. Once inside the cell, this purified phage DNA was able to multiply and produce several hundred new phages. Since the phages that are produced were complete (contain DNA and protein), the phage DNA must contain the genetic information for the entire phage. This interesting experiment is similar to the processes of both transformation and infection and has been termed *transfection*.

The statement that DNA is the genetic material is not universally true. It has been shown that a few viruses contain as their nucleic acid component RNA instead of DNA. In these cases, the genetic material has been shown to be RNA.

In higher animals and plants a large body of circumstantial evidence indicates that here too the genetic material is DNA. Much of this evidence will be discussed in subsequent sections as it relates to specific biochemical processes. Some of the more significant data are as follows:

1. *Localization.* All cells contain DNA which is localized primarily[1] on chromosomes in the nucleus of the cell. This is consistent with the fact that the genes are also located specifically on chromosomes.

2. *Quantity.* Within the same species the amount of DNA per (diploid) cell is the same in all organs. The (haploid) cells such as spermatozoa which contain one-half the amount of genetic capability also contain one-half the amount of DNA.

3. *Stability.* Most of the macromolecules of the cell are constantly being broken down and synthesized. If this were to happen to a gene, valuable hereditary information would invariably be lost. Of all the macromolecules in the cell, DNA is the most metabolically stable.

4. *Sensitivity to chemical and physical agents.* All agents which cause genes to be damaged (mutation) have also been shown to induce changes in the structure of DNA. For example, one such mutagenic agent, ultraviolet light, is capable of producing breaks in the DNA molecule.

In summary, then, all evidence indicates that the genetic material is nucleic acid. In higher animals and plants, bacteria, and most viruses the nucleic acid that carries the genetic information is DNA. In the few viruses that are devoid of DNA, the genetic material is RNA.

5·4 DNA: Chemical structure

Prerequisites to an understanding of how the genes function in the transmission and expression of genetic information is a knowledge of the chemical structure of DNA. Rather than simply presenting our current concept on the

---

[1] Recently, a small amount of DNA has been found outside the nucleus—in the mitochondria of animals and chloroplasts of plants The origin and function of this DNA are, at present, uncertain.

structure of DNA, the observations and experiments which led to this concept will first be offered. In this way the student can obtain some insight into how a structure as complicated as a DNA molecule can be elucidated. Generally, the determination of the structure of a naturally occurring macromolecule involves the following steps.

1. Isolation and purification of the substance.
2. Determination of its component parts.
3. Clarification of how the parts are joined.
4. Determination of the three-dimensional structure of the molecule.

Isolation and purification

In the same decade that Pasteur presented his swan-neck flask to the world, that Mendel reported his genetic studies on garden peas, and that Charles Darwin published the *Origin of Species*, Friedrich Miescher (1844–1895), a student of the German chemist Felix Hoppe-Seyler, became interested in the chemistry of the nucleus. He chose to work with white blood cells because they contained a large and easily observable nucleus. A hospital at Tubingen supplied him with surgical bandages which had been peeled off purulent wounds. From the pus on the bandages he obtained the white blood cells which he then treated with gastric juice (we now know that gastric juice contains an enzyme, pepsin, which digests protein). He observed microscopically that only the shrunken nuclei were left after the pepsin treatment; the remainder of the cell had dissolved. These nuclei were then analyzed and found to have a different composition from any cellular component then known. Since the material was isolated from nuclei, it was called *nuclein*.

Miescher continued his study of nuclein when he returned to his native Swiss city, Basel. He soon found that a more convenient source of this material was salmon sperm. The sperm is exceedingly rich in nuclein. With this favorable material he was able to purify further the nuclein, and when all protein was finally removed, it became clear that the new material was an acid. It was then referred to as *nucleic acid*.

Determination of component parts

Once the nucleic acid molecule could be isolated in pure form, the next task was an analysis of its component parts. The nucleic acid was suspended in a solution of strong acid and then heated in order to break the large molecule into a mixture of many small pieces. The small pieces were then separated and identified. By 1930 each of the component parts had been characterized. Salmon sperm nucleic acid was composed of one part sugar, one part phosphoric acid, and approximately one-fourth part each of the nitrogenous bases adenine, guanine, thymine, and cytosine (Figure 5.1). When the sugar was identified as deoxyribose, the nucleic acid came to be called *deoxyribonucleic acid*, or DNA.

Manner in which the component parts are linked

An intact DNA molecule is a giant structure of great complexity. Even the simplest phage DNA molecule contains more than 5000 submolecules each of deoxyribose and phosphoric acid, and over 1250 each of adenine, guanine, cytosine, and thymine. How are they joined together? Two organic chemists,

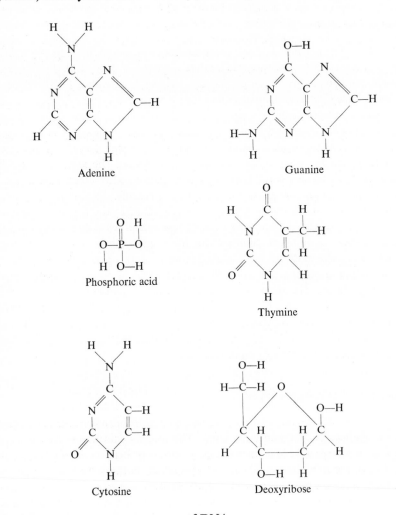

FIGURE 5.1 The component parts of DNA.

Phoebus A. Levene at the Rockefeller Institute and Lord Todd[2] in Cambridge, England, were most instrumental in demonstrating that the components of DNA were joined together to form a long chain of alternating deoxyribose and phosphoric acid units with side chains of the nitrogenous bases (Figure 5.2). An interesting feature of this structure is that although the phosphate-sugar chain is perfectly regular, the molecule as a whole need not be, because the order of bases on the sugar units can vary.

5·5 The three-dimensional structure of DNA

Although the elucidation of the chemical structure of DNA was one of the major achievements of the first half of this century, it was not until its three-dimensional structure became known that there was any clue as to how it might function as the genetic material. The determination of the three-dimensional structure of DNA required data that were obtained with two new techniques, paper chromatography and x-ray crystallography.

Paper chromatography is in practice an exceedingly simple procedure. A small drop of material is applied about 2 inches from the edge of a rectangular sheet of filter paper and allowed to dry there. The edge of the paper is then dipped in an appropriate liquid that slowly moves up the paper by capillary action. As the liquid passes over the point of application the material is pulled along by the liquid at a speed which is characteristic of that substance. By using this technique, a mixture of substances applied to the same

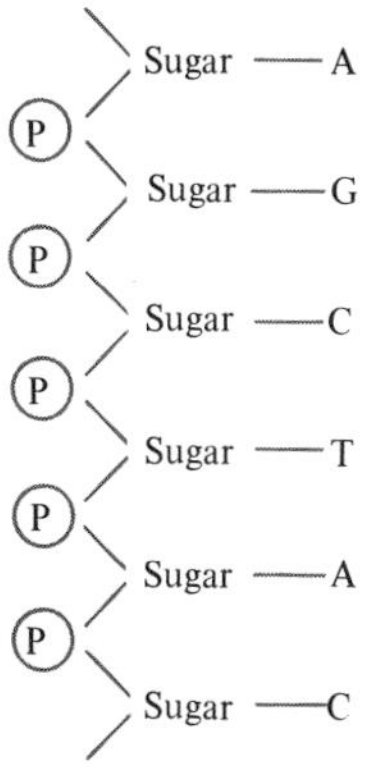

FIGURE 5.2 The chemical structure of DNA. The sugar is deoxyribose, P stands for the phosphate bridge connecting the sugars, A for adenine, G for guanine, C for cytosine, and T for thymine. The molecule must be imagined to extend a very great distance above and below this figure. For example, the smallest DNA molecules (found in viruses) contain approximately 5000 bases.

[2] For his work on nucleic acids, Todd was awarded the 1957 Nobel Prize in Chemistry.

spot can be separated from each other in a few hours. For example, if a mixture of the four nitrogenous bases found in DNA is applied on one spot, and butyl alcohol is allowed to pass over it, the four components will separate from each other in approximately 24 hours (Figure 5.3). The bases can be detected readily on the paper since they all absorb ultraviolet light. In fact, the precise amount of each base can be determined by how much ultraviolet light is absorbed.

Using this simple, but sensitive, technique of paper chromatography, Professor Erwin Chargaff and his collaborators at Columbia University analyzed the base composition of DNA from various sources. By the early 1950s enough data had accumulated to draw some very interesting conclusions. What would you deduce from the data in Table 5.4?

TABLE 5.4

THE BASE COMPOSITION OF DNA OF SEVERAL SPECIES[a]

SOURCE OF DNA	ADENINE	GUANINE	CYTOSINE	THYMINE
Salmon sperm	29.7	20.8	20.4	29.1
Rickettsia (bacteria)	33.8	16.7	15.9	33.6
E. coli (bacteria)	25.2	25.0	25.5	24.3
Yeast	31.7	18.1	17.6	32.6
Micrococcus (bacteria)	14.5	36.1	35.9	13.5
Phage T_2	31.9	18.4	17.9	31.8
Calf thymus	29.8	20.4	20.7	29.1
Calf thyroid	29.6	20.8	20.7	29.1
Calf spleen	29.6	20.4	20.8	29.2

[a] These results are expressed as *mole percent*, which is the number of molecules of one base divided by the total number of molecules of all four bases, multiplied by 100. The data are accurate to ±1 percent.

Conclusion One. The base composition of the DNA is characteristic of the species, differing in composition for different species, but not for the different tissues of any one species.

Conclusion Two. For all DNA samples the amount of adenine equals thymine (A = T) and cytosine equals guanine (C = G). It follows then that A + G (purines) = T + C (pyrimidines).

These conclusions are consistent with the concept of DNA as the genetic material. Since the different species contain different hereditary determinants, it is to be expected that their base composition would also be different. Different tissues of the same species, however, must have an identical base composition since cells divide in such a way that each daughter cell gets an identical copy of genes.

The second conclusion, that A = T and C = G for all species of DNA, came as a complete surprise. You will remember that the purines, A and G,

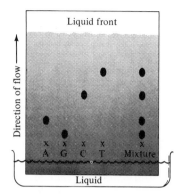

FIGURE 5.3 Paper chromatography of the nitrogenous bases of DNA. Each of the four bases in addition to a mixture was applied at the position marked with an x. As the liquid nears the top of the paper, the mixture becomes resolved, as viewed with an ultraviolet lamp.

are double rings, and the pyrimidines, C and T, are single rings; so there is an equality of large and small rings. Although the significance of these regularities was not at first appreciated, they were soon to provide a major clue in unraveling the three-dimensional structure of DNA.

The next evidence came from x-ray diffraction images of DNA molecules taken by two British workers, Maurice Wilkins and his collaborator, Rosalind Franklin. Because of their short wavelengths, x rays can be used to examine the fine details of molecules. Although the theoretical aspects of x-ray diffraction need not concern us here, what is important is that the pictures were consistent with the following points.

1. DNA from different species give identical x-ray patterns, despite the fact that their base composition varies.

2. DNA molecules are shaped like spaghetti—long and thin (see Figure 2.41); the actual length of the molecule is greater than 30,000 angstroms, whereas it is only about 20 angstroms thick (1 angstrom = one 100-millionth of a centimeter).

3. DNA has a repeating structure every 34 angstroms; that is, as you proceed along the length of the molecule, at regular intervals of 34 angstroms, there is a repetition of the structure.

These, then, were the pertinent experimental facts: the detailed chemical structure of DNA, Chargaff's base pairing rules and Wilkins' x-ray diffraction photographs. There was only one more ingredient that was necessary for the solution of the three-dimensional structure. That ingredient was genius, and it was applied by James D. Watson and Francis H. Crick.

James Watson was a young American biologist at the time (1953) who, having become disillusioned with classical biology, decided to study the structure of macromolecules. After some initial training with Professor Salvador Luria at Indiana University, Watson traveled to Europe, ending up at the Cavendish Laboratories in Cambridge, England. There he met the physicist Francis Crick, who had turned to molecular biology after spending the war years designing mines for the British Admiralty. Working together, they succeeded in a remarkably short time in building a scale model of DNA which fit all of the experimental data.[3] How they arrived at this model is discussed in great detail in Watson's popular book, *The Double Helix*. Of great conceptual value was the discovery two years earlier by Linus Pauling at the California Institute of Technology that protein can exist in the form of a helix.

Figure 5.4 shows the Watson-Crick model for DNA. The most significant feature of the model is that it consists of two strands held together by the attractive forces of the nitrogenous bases. When Watson and Crick constructed scale models of the double-stranded molecule, they soon dis-

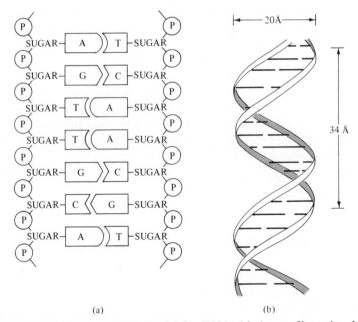

(a)   (b)

FIGURE 5.4 The Watson-Crick model for DNA. (a) A two-dimensional representation of a segment of double stranded DNA. (b) The three-dimensional structure can be visualized as two strands wound around each other to form a double helix, the horizontal bars being the pairs of bases holding the chain together.

[3] For this work on the three-dimensional structure of DNA, Wilkins, Watson, and Crick shared the 1962 Nobel Prize in Medicine and Physiology.

covered that the base pairing must be specific; if there was adenine on one strand, there must be a corresponding thymine on the other strand; likewise, if guanine was on one, cytosine must be on the other. If they attempted to place two of the double ring structures, adenine and guanine, opposite each other, it resulted in a bulge in the molecule and forced the strands apart. Two single ring bases could not be brought close enough together to attract each other. Finally, adenine could not lie opposite cytosine nor could guanine lie opposite thymine because of the geometry of the molecules. Thus *A must always pair with T and G with C.*

Furthermore, in order to maintain the pairing throughout the molecule, it was necessary to twist the two sugar-phosphate backbones in the form of a helical staircase (Figure 5.4b). The steps of the staircase would then be the DNA base pairs. When accurate measurements were made on scale models of this double helix, the diameter turned out to be exactly 20 angstroms and the length for a complete turn was 34 angstroms. There were ten bases on each chain for every complete turn of the helix. The distance between bases on the same chain was thus 3.4 angstroms.

As we have shown, the solution of the three-dimensional structure of DNA had a rather long history. Like almost every discovery, it cannot be attributed to any one person. Starting with Miescher, someone found a bit here, another a bit there, and thus onward until a couple of theoreticians put the bits together and made the breakthrough.

> Science, like the Mississippi, begins in a tiny rivulet in the distant forest. Gradually other streams swell its volume. And the roaring river that bursts the dikes is formed from countless sources.
>
> A. FLEXNER

5·6 Replication of DNA: The Meselson-Stahl Experiment

The Watson-Crick model revolutionized biology and biochemistry not so much because it explained a large body of data on the structure of DNA but because it immediately suggested (to Watson and Crick) how DNA might create an identical copy of itself. Since the structure of DNA consists of two complementary strands, either chain thereby carries the information for making the entire molecule. The mechanism that Watson and Crick proposed for DNA replication is elegant in its simplicity: When ready to replicate, *the two strands separate and each strand serves as a template for the synthesis of its partner.*

Consider a segment of DNA consisting of six bases on each strand (Figure 5.5). When the strands separate, each of the original strands acts as a mold or template for the production of a new chain. The original ACGATT

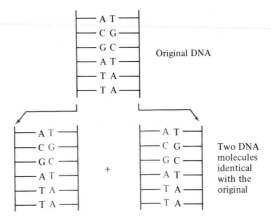

FIGURE 5.5 Watson-Crick model for DNA replication.

strand is copied to yield a fresh strand which must have the sequence TGCTAA because of the rigorous base-pairing requirements imposed by the DNA structure. Likewise, the original TGCTAA strand acts as a template for a new strand with the sequence ACGATT. In this way two complete sets of hereditary information can be produced, each an exact replica of the original.

The Watson-Crick hypothesis for the replication of DNA was put to the test in 1958 by an ingenious experiment designed and performed by two young scientists at the California Institute of Technology, Mathew Meselson and Frank Stahl. Meselson is a physical biochemist who, as a graduate student, had developed a technique for separating DNA molecules according to their densities. He found that if certain salt solutions are centrifuged at very high speeds for about 48 hours, the salt molecules are forced toward the bottom of the tube because of the centrifugal force. Since salt is heavier than water, the higher concentrations of salt toward the bottom of the tube results in the formation of a gradient of densities, the highest density being at the bottom of the tube and the lowest at the top. Furthermore, if DNA is added to the salt solution and then centrifuged, the large DNA molecules collect in a thin band at that part of the gradient which has precisely the same density as the DNA. If a DNA molecule finds itself at too high a density, it floats until it reaches the correct density; if it is at too low a density (that is, too high in the tube), it will sink. Thus DNAs which have different densities will concentrate at different positions in the tube and can thereby be separated.

Stahl as a geneticist provided the biological knowhow for the crucial experiment. They began by growing the bacteria for several days in a medium which contained the heavy isotope of nitrogen, ^{15}N. As far as the bacteria are concerned, the presence of ^{15}N instead of the usual ^{14}N makes very little difference. The bacteria grow and multiply at a normal rate. However, when the DNA is extracted from these bacteria, the nitrogenous bases contain the heavier ^{15}N atoms; thus the DNA is of greater density than normal ^{14}N DNA.

After the bacteria had become completely labeled with ^{15}N, they were transferred to a medium which contained ^{14}N and allowed to continue their growth. Henceforth all DNA would be made with the ^{14}N isotope only! At various times after the transfer to the ^{14}N medium, samples of the culture were removed, and the DNA was extracted and mixed with salt and centrifuged at high speeds. The results are shown in Figure 5.6. The bacteria that were used for these experiments divide every 30 minutes under the conditions

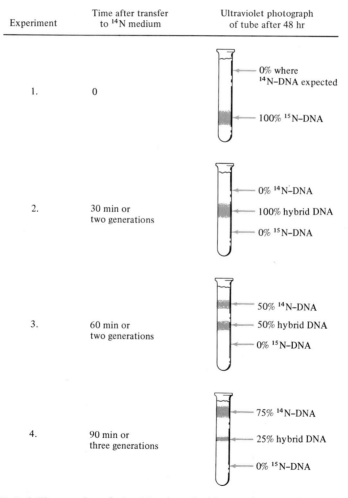

Experiment	Time after transfer to ^{14}N medium	Ultraviolet photograph of tube after 48 hr
1.	0	0% where ^{14}N–DNA expected 100% ^{15}N–DNA
2.	30 min or two generations	0% ^{14}N–DNA 100% hybrid DNA 0% ^{15}N–DNA
3.	60 min or two generations	50% ^{14}N–DNA 50% hybrid DNA 0% ^{15}N–DNA
4.	90 min or three generations	75% ^{14}N–DNA 25% hybrid DNA 0% ^{15}N–DNA

FIGURE 5.6 The results of the Meselson-Stahl experiment. The methods are described in the text.

employed, so that 30, 60, and 90 minutes correspond to 1, 2, and 3 generations, respectively.

Before transfer to the ^{14}N medium (experiment 1), all of the DNA concentrated at a position near the bottom of the tube, corresponding to ^{15}N DNA. After one generation (experiment 2), all of the DNA had a density that was intermediate between ^{15}N and ^{14}N DNA, which will be referred to as hybrid DNA. After two generations (experiment 3), the DNA was 50 percent hybrid and 50 percent ^{14}N DNA. After three generations (experiment 4), the DNA was 25 percent hybrid and 75 percent ^{14}N DNA.

The Watson-Crick replication hypothesis predicts precisely the results of the Meselson-Stahl experiment (Figure 5.7). At the first replication, the ^{15}N DNA would separate into two ^{15}N strands. With each of these acting as a template, two new strands would be constructed; they would contain only ^{14}N so that the first generation would consist of one ^{15}N and one ^{14}N strand; that is, all double strands would be hybrid. With the second replication the strands would again separate. This time, however, half of the strands acting as templates would be ^{15}N and half would be ^{14}N. The new strands would, of course, be only ^{14}N. Thus the DNA would be of two classes, half containing two ^{14}N strands (^{14}N DNA) and half containing one ^{14}N strand and one ^{15}N strand (hybrid DNA). In each successive generation the percentage of hybrid DNA would decrease by one-half.

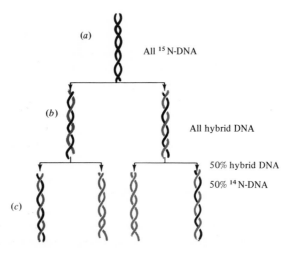

(a) All ^{15}N-DNA

(b) All hybrid DNA

50% hybrid DNA
50% ^{14}N-DNA

(c)

FIGURE 5.7 Interpretation of the Meselson-Stahl experiment. Successive generations in the replication of DNA: (a) Parent DNA with two ^{15}N strands. (b) First-generation, each with one ^{15}N and one ^{14}N strand. (c) Second-generation, two hybrid molecules and two all ^{14}N DNAS.

The Meselson-Stahl experiment does *not prove* the Watson-Crick replication hypotheis. It does make the hypothesis more acceptable, and more importantly, it eliminates several alternative hypotheses which are incompatible with the data. For example, prior to the Meselson-Stahl experiment, it was proposed that the entire double-stranded DNA molecule acts as a template for new DNA. This hypothesis predicts that after one generation in ^{14}N medium, half of the DNA should be the original conserved ^{15}N DNA and half should be composed of two new ^{14}N strands. Since all of the first generation DNA was found to be hybrid, this alternative hypothesis is not tenable and has been discarded.

In addition to the Meselson-Stahl experiment, the Watson-Crick replication hypothesis has been tested in several different ways and in a number of diverse organisms. To date their hypothesis is consistent with all the data and is generally applicable to the replication of DNA of all living organisms, from the simplest viruses to the most complex plants and animals.

5•7   The biosynthesis of DNA

The experiments described thus far in this chapter indicate that the sequence of bases in new chains of DNA is specified by the parental DNA; the information comes from the "old" DNA. But precisely how are the new chains assembled and where does the energy come from?

While work was proceeding on the structure of DNA and the mode by which it replicates in the cell (in vivo), a group at Washington University in St. Louis under the direction of Arthur Kornberg was examining the possibility of replicating DNA in vitro (that is, outside the living cell). They reasoned that the minimum requirements for the biosynthesis of DNA molecules were (1) the component parts, that is, deoxyribose, phosphate, and the four bases, (2) energy to assemble the component parts, (3) DNA to act as a template, and (4) a catalyst to speed up the process.

The component parts could be synthesized in the laboratory or isolated from organic matter, but this would take months. Fortunately, there are several biochemical supply houses which store and sell a large number of research chemicals. To facilitate the experiments, Kornberg purchased the four bases, each of which was already connected to deoxyribose. These molecules which contain one base and one deoxyribose are called *deoxyribonucleosides*. The four deoxyribonucleosides they used in their experiments can be abbreviated dT, dC, dG, and dA. As discussed in Chapter 3 the source of energy for almost all biological reactions is ATP. Thus as a source of energy Kornberg used commercially available ATP. Items (3) and (4) were supplied by extracts of bacteria. The extracts were prepared by grinding the bacteria with powdered glass in a mortar with a pestle. This process disrupts the bacteria and releases the DNA and enzymes of the cell. This, then, was how they obtained the four ingredients for their initial experiments.

The next question was how to measure the newly synthesized DNA. Initially they expected to make very little DNA, if at all; thus they needed an extremely sensitive test for DNA synthesis. The method they chose made use of radioactive tracers. In order to understand this method it is essential to realize that deoxyribonucleosides are soluble in cold acid, whereas DNA is insoluble. One of the deoxyribonucleosides used in the incubation mixture was made radioactive. The test for DNA synthesis was the conversion of acid-soluble radioactivity for example, (dT) to an acid-insoluble form (DNA).

The initial Kornberg experiment consisted of incubating dG, dC, dA, dT (this one being radioactive), and ATP with bacterial extracts (Figure 5.8). At timed intervals, cold acid was added and the acid-insoluble radioactivity was measured in a Geiger counter. Since acid-insoluble radioactivity was found at the later incubation times, the experiment indicated for the first time that *DNA could be synthesized in a test tube.*

For the next ten years, Kornberg and his associates (now at Stanford University) performed a series of brilliant experiments which demonstrated in detail how the DNA was synthesized.[4] The sequence of reactions shown in Figure 5.9 not only demonstrates how DNA is made but more significantly, illustrates the following general principles in the biosynthesis of all macromolecules.

1. *The synthesis of a macromolecule is a multistep process.* Starting with the deoxyribonucleosides, three steps are required to form each of the deoxyribonucleoside triphosphates and an additional step to polymerize them, a total of 13 different reactions.

2. *A different specific enzyme is needed for each step.* The 13 enzymes indicated in Figure 5.8 have been separated and purified. Enzyme 13, which unites the four triphosphates, is called DNA polymerase.

dG

dC

dA   +   ATP   +   Bacterial   ⟶   DNA
                                extract   (radioactive)
                                (DNA + enzymes)

dT (radioactive)

FIGURE 5.8 The first experiment showing the synthesis of DNA in vitro. The four deoxyribonucleosides, dG, dC, dA, dT, were mixed with ATP and a bacterial extract. The newly synthesized DNA was measured by the formation of acid-insoluble radioactivity. At the beginning of the experiment, all of the radioactivity was acid soluble (dT); as the dT became part of the new DNA, the radioactivity became insoluble in cold acid.

[4] For his work on DNA synthesis, Kornberg shared the 1959 Nobel Prize in Physiology and Medicine.

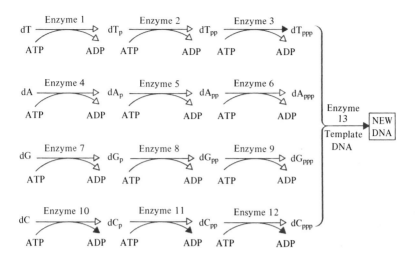

FIGURE 5.9 The sequence of events in the biosynthesis of DNA. To deoxythymidine (dT) is added first one phosphate (dTp), then a second (dTpp), and finally a third (dTppp). In the same manner the other three deoxyribonucleosides are converted into triphosphates. Each reaction is catalyzed by a different enzyme. The source of the added phosphates is ATP which is converted into ADP. The four triphosphates are then joined together by enzyme 13 (DNA polymerase) to form a new chain of DNA whose order is specified by the template DNA.

3. *No ATP is required for the final step.* The four deoxyribonucleoside triphosphates are similar to ATP in that they are "high-energy" compounds. As they are linked together, the terminal two phosphates are released. The energy released by this breakdown is conserved in the formation of new DNA.

4. *The information for the sequence of assembly of the building blocks in the macromolecule is inherent in the template.* Since the sequence of the units which comprise the macromolecule determines its properties and since the properties of the macromolecule are genetically determined, the sequence of the assembled bases must be determined (either directly or indirectly) by the genes (DNA).

In the case of DNA biosynthesis Kornberg clearly demonstrated that the base composition of the newly synthesized DNA is specified by the template DNA and not by the enzyme DNA polymerase or the deoxyribonucleoside triphosphates, which are the building blocks of DNA. Evidence for this conclusion is summarized in Table 5.5. In the first experiment both the DNA polymerase and template DNA came from the bacterium, *E. coli.* The newly synthesized DNA had, of course, the base composition of *E. coli.* Experiment 2 demonstrates that all four triphosphates, DNA polymerase, and the template DNA are absolutely essential for the synthesis of

TABLE 5.5

THE ENZYMATIC SYNTHESIS OF DNA IN VITRO

EXPERIMENT	COMPONENTS IN THE INCUBATION MIXTURE	BASE COMPOSITION OF NEWLY SYNTHESIZED DNA[a]
1	Complete system: dATP, dGTP, dCTP, dTTP, DNA polymerase and template DNA from *E. coli*[b]	51% A + T
2	Omit any one or more ingredients of experiment 1	None formed
3	Like experiment 1 except *E. coli* DNA replaced by micrococcus DNA[b]	29% A + T
4	Like experiment 1 except *E. coli* DNA replaced by phage T_2 DNA[b]	63% A + T
5	Like experiment 1 except *E. coli* DNA polymerase replaced by micrococcus DNA polymerase	51% A + T

[a] The composition of newly synthesized DNA is expressed as mole percent adenine plus thymine; the remainder was guanine and cytosine. In all cases A = T and C = G.

[b] The A + T composition of the template DNAs used were as follows (see Table 5.4): *E. coli* 50 percent, micrococcus 28 percent, and phage T_2 63 percent. These values are accurate to ±2 percent.

DNA. If any one of them is omitted, no DNA is made. If the enzyme is obtained from *E. coli*, but the template DNA from a different source, the new DNA has a base composition indistinguishable from the template DNA (experiments 3 and 4). However, if the template is *E. coli* DNA, but the enzyme is obtained from a different source, the new DNA has the base composition of *E. coli*. The conclusion is clear: The DNA product is determined by the template DNA.

5•8 A sequel: Synthesis of biologically active DNA in the test tube

As in many other fields biology is rapidly catching up with science fiction. What follows is a brief report of one such episode.

Kornberg now asked the crucial question: Was the DNA synthesized in vitro an exact replica of the template DNA? In the January 1968 issue to the *Proceedings of the National Academy of Science* came the first report of the in vitro synthesis of biologically active DNA. For this experiment Kornberg collaborated with Robert L. Sinsheimer of the California Institute of Technology and Mehran Goulian of the University of Chicago.

For more than ten years Sinsheimer had been investigating several interesting properties of the small spherical virus ϕX–174 (see Figure 2.38). The DNA of this bacteriophage is unusual in several ways. First of all,

ϕX–174 DNA is the smallest known naturally occurring DNA molecule containing only 5500 bases. Rather than being double stranded the DNA isolated from phage ϕX–174 contains only a single strand of DNA. Furthermore, the DNA is in the form of a closed ring (Figure 2.44). Most important of all, Sinsheimer had discovered that purified ϕX–174 DNA can infect *E. coli* if the cells were pretreated with a specific enzyme that destroyed the rigidity of the cell wall. With the destruction of the cell wall the bacteria assume a spherical shape, called *spheroplasts*. Single-stranded rings of ϕX–174 DNA can enter *E. coli* spheroplasts, multiply, and then lyse the cell with the release of several hundred new complete ϕX–174 bacteriophages.

The techniques used to reproduce ϕX–174 DNA in vitro are shown in Figure 5.10. The synthesis can be considered in three steps. First, the ϕX–174 template was replicated in the presence of DNA polymerase and the four deoxyribonucleoside triphosphates as previously described. In this case, however, one of the triphosphates contained a heavy atom so that the newly synthesized DNA was more dense than the original template. In the second step an enzyme was utilized to join the ends of the new DNA, forming a double-stranded hybrid molecule. Separation of the synthetic DNA from the original template was accomplished by (1) introducing a single break into the hybrid with DNase, (2) heating to force the broken strands off the remaining closed rings, (3) purifying the closed rings by centrifugation. The introduction of a single break into the hybrid molecule was necessary because the two closed rings were interwoven in such a way that they would not come

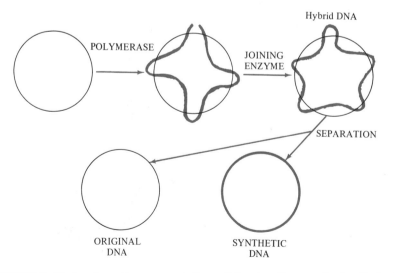

FIGURE 5.10 A schematic representation of the in vitro synthesis of circular ϕX–174 DNA.

apart even when heated. Since the breakage by DNase was random, the same proportion of original and synthetic rings was opened. The remaining closed rings were separated by centrifugation in a salt gradient (see Section 5–6) because the synthetic DNA had a greater density. In this way single-stranded closed rings of φX–174 DNA were synthesized and separated from the starting template.

The synthetic φX–174 DNA was then tested for infectivity with *E. coli* spheroplasts. The DNA made in the test tube was infectious! It penetrated the spheroplasts, multiplied, and then lysed with the release of φX–174 particles that were indistinguishable from the naturally occurring φX–174 phages. This exciting experiment demonstrates that all of the essential genetic information of φX–174 is faithfully reproduced in the synthetic DNA. But does it mean that *life* has been produced (or copied) in a test tube?

QUESTIONS AND PROBLEMS

5.1 What are the similarities and differences between the following:

Chemical structure of DNA and RNA?
Chemical structure of yeast DNA and human DNA?
Transformation and mutation?
Transformation and transfection?

5.2 Outline an experiment which demonstrates that the genetic information of viruses is carried by the nucleic acids.

5.3 What do the following terms signify?

Radioactive isotope	Paper chromatography
DNA polymerase	Transforming principle
DNase	In vitro
Template DNA	Purines and pyrimidines

5.4 List and explain the function of each of the ingredients necessary for the in vitro synthesis of DNA.

5.5 The bacterium, *E. coli*, has a generation time of 30 minutes. Starting with a single radioactive cell and then allowing it to multiply in a nonradioactive environment, how many cells would you have in 2 hours? What fraction of those cells would contain some radioactive DNA? Of those cells which do contain radioactive DNA, what fraction of their DNA is radioactive?

5.6 Would you expect Chargaff's base-pairing rules to be applicable to φX–174 DNA?

5.7 How many twists are there in a phage T_2 DNA molecule (approximately 200,000 bases per molecule)?

SUGGESTED READINGS

Allfrey, V. G., and A. E. Mirsky, "How Cells Make Molecules," *Scientific American*, September 1961 (offprint 92).

Crick, F. H. C., "Nucleic Acids," *Scientific American*, September 1957 (offprint 54). A lucid description of the Watson-Crick model for the structure of DNA.

Hotchkiss, R. D., and E. Weiss, "Transformed Bacteria," *Scientific American*, November 1956 (offprint 18). A popular account of the discovery and early development of transformation.

Kornberg, A., "Biological Synthesis of DNA," *Science*, 131, 1503 (1960). Nobel prize lecture given in Stockholm describing the self-copying of DNA in vitro.

Potter, Van R., *DNA Model Kit*. Minneapolis, Minn.: Burgess Publishing Co., 1959. A paper model of DNA, easy to build.

Tomasz, A., "Cellular Factors in Genetic Transformation," *Scientific American*, January 1969. An up-to-date discussion of the mechanism of genetic transformation.

Watson, J. D., *Molecular Biology of the Gene*. New York: W. A. Benjamin, Inc., 1965. A more advanced treatment of many aspects of this book.

Watson, J. D., *The Double Helix*. New York: Atheneum, 1968. A personal account of the discovery of the structure of DNA. The book was on the best seller list for a number of weeks, attestation to the great interest the public has in modern biology. Many reviews of this controversial book are as interesting as the work itself. (See, for example, P. B. Medawar, *The New York Review of Books*, March 28, 1968.)

6

PROTEIN SYNTHESIS AND THE GENETIC CODE

The frontiers are not east or west, north or south, but wherever a man confronts a fact.

HENRY DAVID THOREAU

In Chapter 5, evidence was presented that indicated that the genetic material is deoxyribonucleic acid (DNA). Furthermore, we described the double helical model of DNA and showed how the model led to a relatively simple scheme for the duplication of DNA. We now turn our attention to the crucial question of how the genetic information is translated by living organisms into observable physiological characteristics. For example, how does a "gene for blue eyes" actually cause an individual to exhibit that characteristic pigmentation? In the last decade there has been a major breakthrough in our understanding of the mechanism by which the genetic information stored in the sequence of bases in DNA exerts its control over the cell. This chapter is an attempt to explain the background for the breakthrough and to discuss our current understanding of gene action.

6·1 The pioneer work of Sir Archibald E. Garrod

In 1902 the English physician Archibald E. Garrod published in *Lancet* the first of a series of papers which dealt with the physiological defect called *alkaptonuria*, This disease, which has been known to medical men for more than 300 years, is characteristically diagnosed by the blackening of the patient's urine on exposure to air. Usually the condition is noticeable in infants from the discoloration of soiled diapers. Garrod began his study by isolating from the urine of patients with alkaptonuria the chemical which turns black on prolonged contact with air. The agent was purified and identified as *homogentisic acid*.

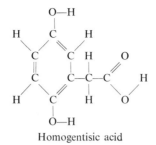

Homogentisic acid

Homogentisic acid is present in the urine of patients with alkaptonuria but absent from the urine of normal individuals. Garrod traced the origin of homogentisic acid to certain foods in the diet, namely the amino acids, tyrosine and phenylalanine. These amino acids are broken down to produce homogentisic acid in all humans. Moreover, normal individuals continue to metabolize homogentisic acid until only carbon dioxide and water are formed. Those afflicted with alkaptonuria, however, cannot metabolize homogentisic acid and therefore excrete it in their urine. The larger the quantity of tyrosine or phenylalanine in their diet, the more homogentisic acid that is excreted.

The diagnosis of alkaptonuria in infants at birth suggested to Garrod that the disease might be inherited. This was supported by examining the marriage records of families having alkaptonuric children. Garrod found that a much larger than expected number were first cousin marriages. Furthermore, examination of the genealogy of some of his patients revealed that alkaptonuria was passed down from parent to child as a simple recessive Mendelian character.

From the data at hand, Garrod conjectured as early as 1908: "We may further conceive that the splitting of the benzene ring in normal metabolism is the work of a special enzyme, that in congenital alkaptonuria this enzyme is wanting." In other words, Garrod was stating for the first time a *relationship between genes and enzymes*. The normal gene results in the production of a specific enzyme, whereas with a defective gene this enzyme is missing.

Garrod reported his results in research papers, books, and even gave a series of lectures to the Royal Society. In spite of these attempts to popularize his theory of gene action, he failed to arouse sufficient interest in biochemists and geneticists to follow through in these studies. It may have been that biochemistry and genetics, both young disciplines at that time, did not have the necessary tools to probe more deeply into this problem. The human organism is a difficult subject to study genetically and biochemically. His life cycle is too long, his offspring too few, and there are severe limitations on subjecting him to chemical analysis.

It may also have been that Garrod, like Mendel, was so far ahead of his time that his contemporaries were not ready to consider seriously his far-reaching gene-enzyme concept.

6·2 The one gene-one enzyme hypothesis

As genetics developed during the first half of this century, certain organisms came to the fore which were well suited for careful genetic and biochemical analyses. One of these was the common bread mold, *Neurospora crassa*. This fungus possesses many advantages for a study on the mechanism of gene action. It can be maintained easily in the laboratory in pure culture; starting with a single cell a genetically homogeneous population can be obtained in a few days. Furthermore, genetic analysis is simplified by the fact that during the greatest part of its life cycle the bread mold has only a single set of genes (haploid) instead of the two sets (diploid) found in higher organisms. Thus complication of dominance and recessiveness is avoided, since genes cannot be hidden by their dominant counterparts. Most important, *Neurospora* can be grown in a minimal medium consisting of water, various inorganic salts, sugar, and one vitamin of the B group, biotin. From these few substances the mold can synthesize all the other components of the cell.

The ease with which both genetic and biochemical studies could be conducted on *Neurospora* induced two scientists at Stanford University, George W. Beadle and Edward L. Tatum, to choose that organism for a careful investigation of the gene-enzyme relationship. As Beadle recollects, "In 1940 we decided to switch from *Drosophila* to *Neurospora*. It came about in the following way: Tatum was giving a course in biochemical genetics, and I attended the lectures. In listening to one of these—or perhaps not listening as I should have been—it suddenly occurred to me that it ought to be possible to reverse the procedure we had been following and instead of attempting to work out the chemistry of known genetic differences we should be able to select mutants in which known chemical reactions were blocked. *Neurospora* was an obvious organism on which to try this approach, for its life cycle and genetics had been worked out by Dodge and by Lindegren, and it probably could be grown in a culture medium of known composition. The idea was to select mutants unable to synthesize known metabolites, such as vitamins and amino acids which could be supplied in the medium. In this way a mutant unable to make a given vitamin could be grown in the presence of that vitamin and classified on the basis of its differential growth response in media lacking or containing it."

The experimental design for the isolation of mutants of *Neurospora* is outlined in Figure 6.1. The same general principles are applicable to many other microorganisms. First, the mold culture is irradiated by x rays or ultraviolet light. This increases the frequency of mutation several thousand fold. Next, a single irradiated cell is placed in a complete medium. If the cell is viable, it multiplies in the complete medium, giving rise to a genetically homogeneous (pure) culture of *Neurospora*. The complete medium contains a wide assortment of nutrients, all of the amino acids, vitamins, purines,

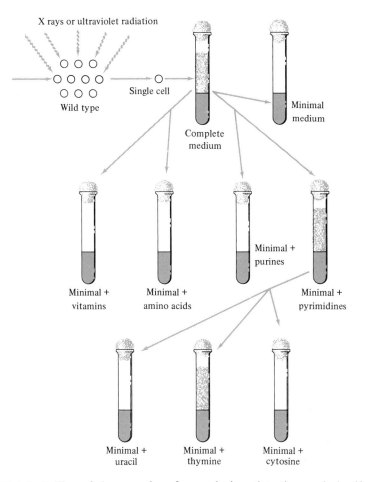

FIGURE 6.1 Outline of the procedure for producing, detecting, and classifying mutations in microorganisms.

pyrimidines, and so on. A sample of the pure culture is then transferred to a minimal medium which contains only sugar, biotin, salts, and water. *Failure of the mold to grow on minimal medium is taken as evidence that a mutation has occurred*. Since the mutant *Neurospora* grows on the complete medium but not on the minimal medium, there must be one or more components in the complete medium which the experimentally produced mutant can no longer synthesize.

The final step is to classify the mutant by identifying the component in the complete medium which is necessary for its growth. This is done by testing the mutant for growth in minimal medium supplemented, in separate cultures, with vitamins, amino acids, or other nutrients. In the example illustrated, the mutant failed to grow when supplemented with vitamins, purines, or amino acids, but grew normally when supplemented with a mixture of the three pyrimidines. By adding the three pyrimidines to the minimal medium one by one, it was determined that the mutant mold had lost the ability to synthesize thymine. The mutant is thus classified as a thymine-requirer.

Using this ingenious technique, Beadle and Tatum isolated a large number of different mutant strains of *Neurospora*, each of which required some nutrient for growth. Each of these mutants was characterized by the loss of the capacity to synthesize a specific nutrient; since all chemical syntheses within the cell are mediated by enzymes, they concluded that the mutational change resulted in a loss of some enzyme(s) involved in the biosynthesis of the nutrient in question. Genes must somehow give rise to specific enzyme(s). This was, of course, what Garrod had suggested (with less evidence) some 20 years earlier.

The second conclusion of Beadle and Tatum was that a *single* genetic factor was responsible for the production of a *single* enzyme, the so-called *one gene-one enzyme hypothesis*. This conclusion was derived from a careful examination of the data from one class of mutants—those which require the amino acid arginine for growth.

The wild type *Neurospora* normally synthesizes arginine by a sequence of enzymatically catalyzed reactions which can be abbreviated as follows:

$$\text{Minimal medium} \xrightarrow{\quad\quad} A \xrightarrow{\text{Enzyme 1}} B \xrightarrow{\text{Enzyme 2}} C \xrightarrow{\text{Enzyme 3}} \text{Arginine}$$

In the final three steps, a compound which we can refer to as A is transformed into B, then to C, and finally to arginine. Each of these steps is catalyzed by a different enzyme. Beadle and Tatum reasoned that some of their arginine-requiring mutants might be lacking enzyme 1, others enzyme 2, and still others enzyme 3. To test this possibility they investigated the growth of their mutants on minimal medium plus compound A, B, C, or arginine (Table 6.1).

The arginine-requiring mutants fell into four classes (I–IV) according to their growth requirements. Class I mutants grew only if supplemented with arginine, II grew on C or arginine, III grew on B, C, or arginine, and IV grew on A, B, C, or arginine. In no case could a mutant grow on A or B but not C. These data are logically consistent with the following interpretation: Class I mutants lack enzyme 3, II lack enzyme 2, III lack enzyme 1, and IV lack a necessary enzyme for the production of compound A. Consider, for example, Class III mutants. Since they are able to grow on minimal plus B, they must contain enzymes 2 and 3 which are necessary for the conversion of B into arginine. However, since they cannot grow on A, Class III mutants must

TABLE 6.1

GROWTH REQUIREMENTS OF FOUR CLASSES OF ARGININE-REQUIRING MUTANTS OF *NEUROSPORA*

ARGININE-REQUIRING MUTANT	GROWTH[a] ON MINIMAL PLUS				
	NOTHING	A	B	C	ARGININE
I	−	−	−	−	+
II	−	−	−	+	+
III	−	−	+	+	+
IV	−	+	+	+	+

[a] + indicates growth; − indicates no growth.

lack enzyme 1 which catalyzes the conversion of A into B. The same type of reasoning also applies to the other classes.

Two follow-up experiments were then performed which provided additional support for the one gene-one enzyme hypothesis. First, genetic experiments (similar to those described in Chapter 4) with *Neurospora* demonstrated that each of the four classes of arginine-requiring mutants was deficient in a different gene. Second, direct biochemical tests revealed that in each case the appropriate enzyme was indeed missing.

The Beadle and Tatum experiments[1] thus provided conclusive support for the one gene-one enzyme hypothesis. Since that time similar conclusions have emerged from biochemical genetic studies on various other organisms, from bacteria to man. The gene-enzyme relationship now appears to be a general biological phenomenon.

6·3   The mechanism of protein synthesis

During the 1950s molecular biologists realized that genes (DNA) exert their control by the production of specific enzymes (protein). However, it was not at all obvious at the biochemical level *how* DNA could dictate the structure of proteins. In the first place, there was no apparent complementation in chemical structure between DNA and protein as had been demonstrated previously for the two strands of DNA. Furthermore, careful cytochemical analyses had revealed that DNA is located in the nucleus, whereas protein is synthesized in the cytoplasm of cells. How, then, can DNA direct protein synthesis? This paradoxical situation only recently has been resolved. As the details of the mechanism of protein synthesis have been unraveled, the indirect, but controlling role by DNA on protein synthesis has emerged.

[1] For their experiments on the gene-enzyme relationship in *Neurospora*, Beadle and Tatum received a share of the 1958 Nobel Prize for Medicine and Physiology.

In 1953 Paul C. Zamecnik and co-workers at Harvard University reported for the first time the synthesis of protein in vitro. Their experiment consisted of mixing the following four ingredients in a test tube: (1) ATP; (2) 20 amino acids, one of them radioactive; (3) cellular particles called ribosomes (see Figure 2.21), and (4) soluble extract of cells. After 30 minutes of incubation, the reaction mixture was chilled and cold acid was added to it. The measurement of protein synthesis was based on the fact that the amino acids are soluble in acid, whereas proteins are insoluble. The reaction mixture was then centrifuged to separate the soluble and insoluble fractions. Zamecnik and his associates observed radioactivity in the acid-insoluble fraction and thus concluded that protein was synthesized during the experiment. From careful studies on the role of each of the four components in the synthetic process came our current understanding of protein synthesis.

The first step in protein synthesis is the "activation" of the amino acids by ATP:

$$(1) \qquad ATP + amino\ acid \xrightarrow{\text{Enzyme}} Amino\ acid\text{-}AMP + P\text{---}P$$

In this reaction the ATP and amino acid are combined with the release of pyrophosphate (P—P) from ATP. The energy lost in splitting off the phosphates is conserved in the formation of the amino acid-AMP complex. For each of the 20 amino acids there is a different enzyme that catalyzes the activation. The source of these enzymes in Zamecnik's concoction was the soluble extract.

The second step in protein synthesis was discovered quite by accident. One of Zamecnik's associates, Mahlon B. Hoagland, considered the possibility that RNA as well as protein was being synthesized in their reaction mixture. To test this hypothesis he added to the protein-synthesizing system one additional ingredient, radioactive uracil. As mentioned previously uracil is a component of RNA but not DNA. If RNA were synthesized, the radioactive uracil would also be found in the acid-insoluble fraction. Protein and RNA could then be distinguished, since RNA is soluble in *hot* acid, whereas protein remains insoluble.

The experiment was performed. The first result Hoagland obtained was that the cold acid-insoluble fraction was radioactive. This could be due to amino acids entering into proteins and uracil entering RNA or both. Next, the acid-insoluble fraction was suspended in acid and heated for 1 hour. As expected radioactivity was found in the hot acid-insoluble fraction, indicating that protein was synthesized. The exciting result was that the hot acid-soluble fraction was also radioactive. The tentative conclusion was that RNA had also been synthesized. In one of Hoagland's several controls he left out the radioactive uracil although, of course, the radioactive amino acid was still present. Nevertheless, he still found radioactivity in the hot acid-soluble fraction containing the RNA. The logical conclusion was that some of the radioactive amino acid had become attached to RNA and was thus rendered soluble in hot acid. These experiments are summarized in Figure 6.2.

The demonstration of an amino acid-RNA complex led to an understanding of the second step in protein synthesis:

$$(2) \quad \text{Amino acid-AMP} + \text{RNA} \xrightarrow{\text{Enzyme}} \text{Amino acid-RNA} + \text{AMP}$$

In this reaction an amino acid is transferred from AMP onto an RNA molecule. This type of RNA is referred to alternatively as transfer RNA (tRNA) or soluble RNA (sRNA). The tRNAs, containing approximately 80 nucleotides, are the smallest naturally occurring nucleic acid molecules. There is at least one specific tRNA for each amino acid and, in the case of some amino acids, there may be as many as five different, specific tRNA molecules. In one of the great achievements of modern chemistry, Robert W. Holley and his co-workers have recently determined the entire sequence of bases in a tRNA for the amino acid alanine.

Reactions (1) and (2) can occur when a soluble extract of cells is mixed with ATP and the amino acids. All the enzymes and tRNAs necessary for the formation of amino acid tRNA complexes are contained in the extract. However, the reaction ceases, and no protein is formed unless cellular particles called *ribosomes* are also added. We can thus express the next reaction of protein synthesis as follows:

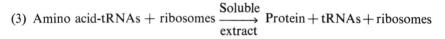

$$(3) \quad \text{Amino acid-tRNAs} + \text{ribosomes} \xrightarrow[\text{extract}]{\text{Soluble}} \text{Protein} + \text{tRNAs} + \text{ribosomes}$$

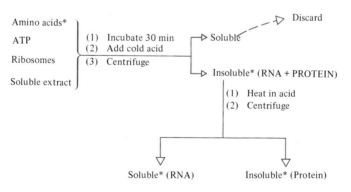

FIGURE 6.2 The formation of an amino acid-RNA complex. The asterisks indicate the presence of radioactivity. The experiment indicated that amino acids become incorporated into both the hot acid-insoluble (protein) and soluble (RNA) fractions.

Investigators in many laboratories throughout the world have studied the ribosome and its reactions. Ribosomes have been found in the cytoplasm of all types of cells, from bacteria to mammals. The ribosome is a spherical or ellipsoid granule, consisting of approximately equal weights of protein and RNA. Ribosomal RNA is much larger than tRNA, containing over 1000 bases per molecule. Each ribosome contains more than 50 different proteins; some of these proteins are enzymes which control the linking together of the amino acids and the release of tRNA.

Several independent experiments have demonstrated that the ribosomes are the site of protein synthesis in living cells. In one such experiment growing bacteria were exposed to radioactive amino acids for only 5 seconds. The bacteria were then chilled with ice to prevent further reactions, and the cells rapidly harvested and disrupted. After separation of the broken cell extract into a soluble fraction and a ribosomal fraction by centrifugation, each was analyzed for acid-insoluble (protein) radioactivity. With this very short labeling period, the radioactivity was associated with the ribosomal fraction, indicating that it was the site at which the protein was manufactured. In a parallel experiment the bacteria were labeled for 5 seconds, but then allowed to incubate with excess nonradioactive amino acids for 2 minutes prior to chilling. Under these conditions no radioactivity was found in the ribosomal fraction. It would appear therefore that protein is first synthesized on the ribosomes, then liberated into the soluble portion of the cell.

Although the discovery of the role of ribosomes in protein synthesis was highly important, it did not solve the fundamental problem of how DNA controls the formation of protein. One approach to this problem was to ask the question: Do the ribosomes determine the *kind* of protein being synthesized? From the pioneer (1948) experiments of Seymour S. Cohen at the University of Pennsylvania, it was known that immediately after phage T_2 infects *E. coli*, synthesis of ribosomes stops. Since the phage coat and other phage specific proteins are produced in large quantities after infection, it follows that *E. coli* ribosomes are nonspecific; prior to infection the ribosomes synthesize *E. coli* proteins, and after infection they synthesize phage proteins. Thus the ribosomes are nonspecific protein assembly factories. The information for specifying the type of protein to be synthesized must reside elsewhere.

About 1960 François Jacob and Jacques Monod at the Pasteur Institute in Paris put forth a suggestion which explained the source of the information-containing component. They proposed that the information for determining the protein structure was transmitted from the genes to the ribosomes by a third type of RNA molecule, which they called messenger RNA (mRNA). According to their model the mRNA is *transcribed* from the DNA template, using base-pairing rules, and would thus have embodied in it the genetic information. Once synthesized, the mRNAs migrate to the cytoplasm, where they become attached to the ribosomes. The message would be "read" at the ribosomes when the various tRNAs, each bearing its specific amino acid, became aligned in a sequence dictated by the mRNA. Finally, when the amino acid-tRNA complex is fixed in position, an enzyme unites the amino acid to its neighbor, with the release of the tRNA. A schematic representation of these processes is shown in Figure 6.3.

Several predictions can be made from the Jacob-Monod model: (1) the size of the mRNA should vary considerably, depending on the size of the protein it codes for; (2) the mRNA should have a base composition similar to DNA (except uracil should replace thymine); (3) the mRNA should be complementary to a portion of one DNA strand; and (4) the mRNA should stimulate protein synthesis in vitro and also determine the type of protein made. Using modern techniques of centrifugation, mRNA recently has been

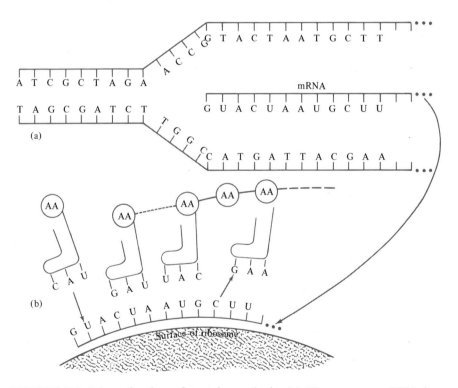

FIGURE 6.3 Schematic view of protein synthesis. (a) The messenger RNA is copied from a segment of DNA, according to the base-pairing rules, except that uracil takes the place of thymine. (b) The mRNA then migrates to the cytoplasm where it becomes attached to a ribosome. The tRNA-amino acid complexes then wait their turn to line up on the mRNA template. When the amino acid (AA) has been joined to the growing protein chain, the "empty" tRNA is released.

isolated from a variety of cells. In all cases the purified mRNA has properties that coincide with those predicted. In addition, it has been shown that in many cells mRNA is metabolically unstable. In bacteria, for example, the mRNA seems to break down as soon as it has done its job of programming a few protein molecules.

The discovery of the role of mRNA in protein synthesis brought together the various bits of genetic and biochemical data, thus forming a unified theory for gene action. The scheme, which is summarized in Figure 6.4, may seem complicated, but it is certainly ingenious.

The following points should be emphasized: (1) all three species of RNA are made from the DNA template; (2) the ribosome is formed from ribosomal RNA and protein, and only becomes programmed for protein synthesis when a mRNA is attached to it; (3) the amino acids enter the programmed ribosome after they have been activated and linked to a specific tRNA. The tRNA acts as a two-handed molecule, one hand holding onto the specific amino acid, the other binding to a specific sequence of bases on the mRNA. In this manner the amino acids are incorporated in the order coded for by the mRNA, which in turn is coded for by the DNA.

6·4 The genetic code

In light of what is presently known about the mechanism of protein synthesis, the one gene-one enzyme concept can be restated as follows: One segment of DNA is responsible for the production of a mRNA which codes for a specific sequence of amino acids. What is this code? Precisely which sequence of bases on the mRNA "spells out" which amino acid?

From a purely theoretical argument we can conclude that it requires more than two bases to code for a single amino acid. Suppose only one base were to be used as a code word for an amino acid. Then the four bases could specify only four amino acids. Since there are 20 different amino acids that are found in proteins, the single base code is excluded.

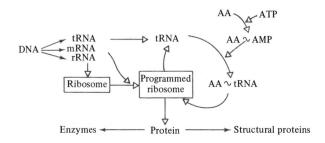

FIGURE 6.4 Summary of reactions involved in protein synthesis.

Using two bases to code for one amino acid yields 4 × 4, or 16, possible combinations.

AC	CC	GC	UC
AA	CA	GA	UA
AG	CG	GG	UG
AU	CU	GU	UU

Since this is still not enough to account for the 20 amino acids, we must conclude that more than two bases code for one amino acid. If three bases are used in each code word, there are 4 × 4 × 4, or 64 possible triplets, which is more than adequate.

Although these simple theoretical considerations were discussed as far back as 1952, no significant progress was made in "cracking the code" until 1961. Then, at the Fifth International Congress of Biochemistry in Moscow, a young biochemist from the National Institutes of Health, Marshall W. Nirenberg, reported his experiments on protein synthesis in vitro. His results amazed the scientific world. He had made the breakthrough which led directly to deciphering the code.

Nirenberg and his associates performed a series of experiments similar to those of Zamecnik, except for one important step: Nirenberg added deoxyribonuclease (DNase) in addition to the ATP, ribosomes, amino acids, and soluble extract. Under these conditions protein synthesis stops after about 20 minutes (Figure 6.5).

Why does the reaction stop? To an empiricist such a question suggests a simple experiment: After 30 minutes with DNase, add back to the incubation mixture in separate tubes, ATP, amino acids, ribosomes, tRNA, and mRNA.

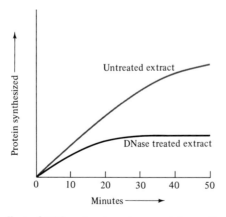

FIGURE 6.5 The effect of DNase treatment on protein synthesis in vitro. Protein was measured by the incorporation of radioactive amino acids into the acid-insoluble fraction.

The results of such an experiment is shown in Figure 6.6. The answer is clear: Only the mRNA stimulates. They reasoned that during the initial 30 minutes, the mRNA was broken down. Since no template DNA was present (DNase destroyed it), the mRNA could not be replenished and protein synthesis ceased.

These experiments paved the way for a comparative study of the effect of various mRNAs on protein synthesis. Once the incubation mixture becomes depleted of the mRNA originally present in the soluble extract, the extent and type of protein synthesized depend entirely on added mRNA. Nirenberg and his associates found, for example, that RNA from virus-infected cells greatly stimulated protein synthesis, whereas RNA isolated from ribosomes stimulated only slightly. Then, either by design or chance (only Nirenberg knows for sure), they decided to test an artificial RNA-like polymer which is called in laboratory jargon "poly-U." Poly-U is a synthetic macromolecule which has a structure identical to that of RNA, except that it contains only one type of base, uracil.

Their experiments revealed that poly-U was a surprisingly efficient messenger for protein synthesis. But if the RNA message was UUUUUUUUU ..., what was the protein product? Chemical analysis revealed that the reaction product was a polymer containing only one amino acid, phenyl-alanine.

Thus Nirenberg and his associate, J. Heinrich Matthaei, concluded, "Addition of poly-U resulted in the incorporation of phenylalanine alone into a protein resembling polyphenylalanine. Poly-U appears to function as a synthetic template, or messenger RNA, in this system. One or more uridylic

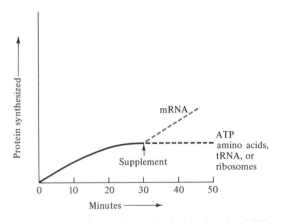

FIGURE 6.6 The stimulation of protein synthesis in vitro by ATP, amino acids, ribosomes, tRNA, and mRNA. The initial conditions were the same as in Figure 6.5 (+DNase). After 30 minutes, the incubation mixture was divided into five portions; each was then supplemented with one of the substances.

acid residues appear to be the code for phenylalanine." Assuming a triplet code, the first word (codon) in the genetic dictionary then becomes UUU = phenylalanine.

With the path clearly prepared by Nirenberg's initial discovery, progress was rapid. Artificial RNA polymers were synthesized containing every possible combination of the four bases. They were then tested for their ability to stimulated protein synthesis in vitro. By analyzing the relative amount and distribution of each amino acid incorporated into protein under the guidance of these synthetic RNAs, it was possible to assign codons to every amino acid. Especially useful were the nucleic acids chemically synthesized by H. G. Khorana and his associates at the University of Wisconsin. The following specific example (Table 6·2) illustrates how some of these assignments were made.

TABLE 6.2

ASSIGNMENT OF CODONS USING SYNTHETIC mRNAs

MESSENGER RNA	AMINO ACID INCORPORATED	CODON
ACACACAC...	Threonine	ACA
	Histidine	CAC
CAACAACAA...	Threonine	ACA
	Glutamic acid	CAA or AAC
	Asparagine	CAA or AAC

An artificial messenger containing an alternating sequence of adenylic and cytidylic acids, ACACACACACAC..., led to the incorporation of two amino acids, threonine and histidine, also in alternating sequence. This is consistent with a triplet code in which one of the amino acids is coded for by ACA, the other by CAC. It was also found that the synthetic mRNA containing the sequence CAACAACAACAACAA... led to the incorporation of three amino acids, glutamate, asparagine, and threonine. Depending on where the reading begins, the possible codons are CAA, AAC, and ACA. Since threonine is the only amino acid incorporated in both these examples and since ACA is the only triplet present in both messengers, then ACA must be the codon for threonine. It follows that CAC codes for histidine.

By the end of 1968 the total assignment of the 64 triplet codons was completed (Table 6.3).[2] From an analysis of the data, several generalizations emerge.

[2] The triplet, R. W. Holley, H. G. Khorana, and M. W. Nirenberg, shared the 1968 Nobel Prize in Medicine and Physiology for their independent but related experiments on the involvement of RNA in protein synthesis.

TABLE 6.3

THE GENETIC CODE[a]

SECOND	U	C	A	G	THIRD
	Phe	Ser	Tyr	Cys	U
U	Phe	Ser	Tyr	Cys	C
	Leu	Ser	"Stop"	"Stop"	A
	Leu	Ser	"Stop"	Trp	G
	Leu	Pro	His	Arg	U
C	Leu	Pro	His	Arg	C
	Leu	Pro	Gln	Arg	A
	Leu	Pro	Gln	Arg	G
	Ileu	Thr	Asn	Ser	U
A	Ileu	Thr	Asn	Ser	C
	Ileu	Thr	Lys	Arg	A
	Met[b]	Thr	Lys	Arg	G
	Val	Ala	Asp	Gly	U
G	Val	Ala	Asp	Gly	C
	Val	Ala	Glu	Gly	A
	Val[b]	Ala	Glu	Gly	G

[a] This table shows the best codon assignments as of January 1969. Standard abbreviations are used for the 20 amino acids. The left-hand column lists the four bases representing the first letter of the codon; the second letter is one of the bases listed across the top, and the third letter is listed vertically in the last column. The codons indicated by "stop" are believed to cause chain termination.

[b] Two codons are believed to be the signals for chain initiation, AUG and GUG, when they occur either at the beginning of the mRNA or following a "stop." The codon AUG specifies methionine both at the beginning of the protein and internally. The codon GUG normally specifies valine; however, when the GUG occurs at the beginning of an mRNA or following a "stop," it codes for methionine.

1. *The code is degenerate.* Degeneracy means that several different triplets can specify the same amino acid. For example, both UUU and UUC code for phenylalanine.

2. *There are codons for start and stop.* The mRNA is not simply read from one end to the other. Instead, there are specific codons which determine where reading begins and ends. Chain growth is initiated by the codons GUG and AUG and is terminated by UAA, UAG, or UGA.

3. *The code is largely, if not entirely, universal.* A given mRNA will be read in the same manner by species at opposite ends of the evolutionary scale. For example, poly-U stimulates phenylalanine incorporation in cell extracts from bacteria to mammals.

Chapters 5 and 6 have concentrated on a molecular approach to genetics, an approach which in the last 20 years has yielded the most extraordinary results in the history of the life sciences. The gene has been identified and chemically characterized; and giant strides have been taken toward an understanding of the mode of gene action. However, it would be wrong to leave the impression that the genetic processes are now understood completely at the molecular level.

Unraveling the genetic code is equivalent to learning the alphabet. We must still learn to read—and to understand what we read. Buried in the sequence of bases in DNA are the genetic histories of species, for only through changes in the order of bases are mutation and subsequent evolution possible. At present no methods are available for determining the sequence of bases in DNA.

Finally, an age-old dream of geneticists is to control mutation in order to eliminate genetic defects and to produce mutants with desirable features. In principle it should be possible to direct mutations, but this is a case where principle and practice are far apart. We hope that by the time techniques of genetic manipulation are available, society will have developed to the point where these discoveries will not be misused.

QUESTIONS AND PROBLEMS

6.1 The wild type *E. coli* can grow on sugar as its sole source of carbon. Five mutants were obtained which required vitamin B_1 in addition to the sugar for growth. When these mutants were tested for growth with precursor molecules (G, H, I, R, and T), the following results were obtained:

ORGANISM	VITAMIN B_1	NOTHING	SUGAR PLUS				
			G	H	I	R	T
Wild type	+	+	+	+	+	+	+
Mutant 1	+	−	+	+	−	−	+
Mutant 2	+	−	−	+	−	−	+
Mutant 3	+	−	+	+	+	−	+
Mutant 4	+	−	−	−	−	−	+
Mutant 5	+	−	−	−	−	−	−

(a) In the formation of vitamin B_1 what is the order of the five precursors?

(b) Where is each mutant blocked?

(c) How does mutant 1 differ from the wild type with respect to the detailed structure of its DNA, RNA, and protein?

6.2 Assume that a particular gene segment of DNA consisted of the tetranucleotide TTTC repeated 30 times; that is, TTTCTTTCTTTCTTTCTTTC(TTTC)$_{25}$.

(a) How large a protein (number of amino acids?) would you predict would result from this strand of DNA?

(b) Using Table 6.3, what would be the amino acid sequence in the resultant protein? (Use standard abbreviations.)

(c) Draw schematically the complex that is formed between the mRNA derived from this gene and the tRNA amino acids. Show the base sequences wherever possible.

6.3 Compare the different types of RNA with respect to their size, base composition, and function in protein synthesis.

6.4 Explain the poem in Section 1–9 with respect to DNA, RNAs, and protein.

6.5 Although DNA is a double helix, we know that only one of the two strands is used to produce protein. Suggest a mechanism that allows one strand to be transcribed but not the other.

SUGGESTED READINGS

Beadle, G. W., "Genes and Chemical Reactions," *Science*, 129, 1715 (1959). The Nobel Prize lecture given in Stockholm.

Borek, E., *The Code of Life*. New York: Columbia University Press, 1965. A nontechnical account of biochemical genetics by the winner of the 1961 Thomas Alva Edison Foundation Award for the best science book for youth.

Nirenberg, N. W., "The Genetic Code: II," *Scientific American*, March 1963.

Nomura, M., "Ribosomes," *Scientific American*, October 1969.

Sinsheimer, R. L., *The Book of Life*. Reading, Mass.: Addison-Wesley Publishing Co., Inc., 1967. An easily read discussion of the nature of the genetic code. Conceptual relations are clearly developed, independent of the specific biochemistry.

Sonneborn, T. M. (ed.), *Control of Human Heredity and Evolution*. New York: Macmillan Company, 1965. An edited record of a symposium on the moral, ethical, and philosophical problems with which society will be faced in the future. Particularly pertinent are the articles by S. E. Luria, E. L. Tatum, and H. J. Muller.

7

GROWTH AND REGULATION*

As biological studies proceed to a molecular and a genetic level, the parallel between mammalian and microbial cells has become increasingly prominent.

BERNARD D. DAVIS

The phenomenon of growth is the most dramatic of all biological events. Growth may be defined as the orderly increase of all cellular constituents, leading to an accurate duplication of the existing pattern. We have left this discussion for the last because growth encompasses all other biological processes such as energy generation, biosynthesis, and reproduction, topics which have been discussed in previous chapters. The study of growth is not a specialized subject or field of research: It is the basis of biology.

In this chapter, we discuss nutritional requirements for growth, the growth cycle, and finally the regulation of expression of genetic material.

7·1 Requirements for growth

The concept of the unity of biochemistry proposes that the basic chemistry of all cells, microbial, animal or plant, is essentially the same. All cells use DNA as their genetic material, possess three types of RNA which are used in the synthesis of proteins, are bounded by membranes which are made up of proteins and lipids, and so on. Although cells are inherently similar in chemical makeup, macromolecules, such as nucleic acids, proteins, and polysaccharides, produced by one cell are not incorporated directly into the cellular structures of other cells. Each cell must therefore synthesize for itself a large variety of cellular constituents. Thus every living creature must find in its environment the required chemical components (the raw materials) and the energy sources for growth.

* This chapter was written by Dr. R. J. Martinez, Department of Bacteriology, University of California, Los Angeles, California.

The varied nutritional needs of different organisms are an excellent illustration of diversity in the biological world. Most animals must be supplied with a complex diet containing, in addition to water and minerals, a large number of amino acids and an array of vitamins, fats, and carbohydrates. The bacterium *Escherichia coli*, on the other hand, can grow on a medium consisting of only water, minerals, and the sugar, glucose—even though its basic cellular chemistry is essentially the same as that of an animal cell. The reason for this diversity in nutritional needs is that many cells, such as *E. coli*, possess all the necessary enzymes for the synthesis of amino acids, purines, pyrimidines, vitamins, fats, and so on; these enzymes are not found in the nutritionally demanding animal cells. It follows that organisms having a wide array of biosynthetic enzymes have less complex nutritional needs than those creatures with restricted biosynthetic abilities.

In essence, diversity in nutritional requirements is a manifestation of evolution. As discussed in Chapter 1 primordial cells presumably arose in "organic soups," rich in a multitude of complex nutrients. We may speculate further that as these requisite compounds were exhausted from the "soup" as a result of their consumption by the cell population, the process of biosynthetic evolution was initiated; that is, by mutation and natural selection organisms arose that contained enzymes whose function was to convert chemicals in rich supply to those that were in demand by the cells. In this way primordial cells gradually developed their biosynthetic machinery and evolved from highly nutritionally demanding creatures with a paucity of biosynthetic enzymes to nutritionally independent organisms rich in biosynthetic properties. Eventually photosynthetic organisms evolved with the capacity to obtain their cellular carbon by fixation of carbon dioxide gas from the atmosphere, their nitrogen by fixation of nitrogen gas, and their energy from sunlight. These creatures represent the most biosynthetically advanced forms of life. In the presence of sunlight they can grow in essentially air and water containing traces of minerals.

With the advent of these highly evolved photosynthetic forms, other cells underwent a gradual loss of biosynthetic powers—a type of retrograde evolution. They obtained their needed nutrients preformed from other cell types. Humans, for example, require eight amino acids for growth. These are obtained by eating animal or plant proteins. Certain disease-causing bacteria and viruses are so nutritionally demanding that they have not yet been grown outside their living host animals.

7·2   The growth curve

The growth of the bacterium *E. coli* has been investigated more extensively than that of any other living form. Table 7.1 lists the minimum requirements for the growth of *E. coli*. Glucose serves as the carbon and energy source

TABLE 7.1

SIMPLE MEDIUM FOR THE GROWTH OF E. coli

Water	1 liter
Glucose	2 grams
Ammonium chloride	1 gram
Magnesium sulfate	0.5 gram
Phosphates	10 grams
Trace elements	0.02 milligram of each

from which the cells manufacture all cellular constituents and provides the energy necessary for growth. *Ammonium chloride* is the source of nitrogen required for the synthesis of proteins, nucleic acids, and other nitrogen-containing components of the cell. The *phosphates* furnish the phosphorus necessary for nucleic acid synthesis and serve to maintain a favorable condition with respect to acidity and alkalinity. *Magnesium sulfate* also has a dual role—sulfur is incorporated into proteins and certain lipids of the cell; magnesium functions as a cofactor for many enzyme reactions. Other elements necessary for bacterial growth are required in such minute amounts that they are usually present as contaminants in the other ingredients or they may be added as "trace elements."

Let us consider the events that take place when a flask containing this simple medium is inoculated with *E. coli*. The culture is incubated on a shaking machine to provide adequate aeration, since oxygen is required by cells for the production of ATP via respiration.

Bacteriologists have devised simple methods for measuring the number of bacteria in a culture. Using these techniques, the numbers of bacterial cells per unit volume of culture fluid are measured as a function of time of incubation. A curve similar to that shown in Figure 7.1 is usually obtained. This plot of the number of cells as a function of time, termed the growth curve,

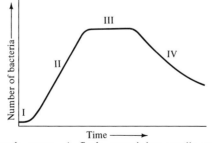

FIGURE 7.1 The growth curve. A flask containing sterile nutrient medium is inoculated at time 0. Samples of the culture are withdrawn at intervals, and the number of cells per unit volume of culture is plotted against time. Four phases of growth are usually observed: I, lag phase; II, exponential growth phase; III, stationary phase; IV, death phase.

is characteristic of the growth of virtually all populations, whether viruses, bacteria, rabbits, or elephants. Only the rates will vary.

The growth curve may be divided conveniently into four phases or periods: (I) *lag phase*, (II) *exponential phase*, (III) *stationary phase*, and (IV) *death phase*. The lag phase, which occurs immediately after the inoculum is introduced into a new medium, is a period of stationary population. Following a short period of increasing growth rate, the culture enters the exponential phase, sometimes called the logarithmic growth phase. During this period all viable cells are dividing at the maximum rate. This phase is followed by a short period of decreasing multiplication rate, and the culture then enters into the stationary phase. The last phase of the growth curve is called the death phase.

The lag phase

The lag phase is a period of adjustment. It is observed when a fresh medium is inoculated with cells from an old culture, or when the environmental conditions, for example, the composition of the medium or the temperature of growth are changed. In the early 1920s the microbiologist A. T. Henrici noted that bacterial cells increased their size two to three times when introduced into a fresh medium. Later it was shown that cells in lag phase synthesize nucleic acids and proteins. It appears, then, that there is a greater delay in cell division than in actual growth as represented by the synthesis of cell material. Why, then, if cells are undergoing net synthesis of nucleic acids and proteins, is there a lag in cell division? Old cells that are no longer dividing may be deficient in certain substances that are critical for the process of cell division. These substances must be synthesized and properly balanced with the other cellular constituents before division takes place. As an illustration, recently Moselio Schaechter at the University of Florida demonstrated that old bacteria, that is, cells in the stationary phase of growth, have fewer ribosomes than cells in the exponential phase. These cells must therefore synthesize ribosomes before entering into exponential growth. *The lag phase constitutes a period of adjustment for initiation of exponential growth.*

It follows that cells that are growing exponentially, and are therefore fully adjusted for rapid cell division, should not show a lag when transferred to fresh medium of the same composition. The lag phase can be completely eliminated by using as an inoculum a culture from the exponential phase. The lag phase may last anywhere from minutes to many hours to even days depending on the size and age of the inoculum.

The exponential phase of growth

The exponential growth phase is that period during which the rate of increase of cells is maximum and constant. Every organism has a rate of growth during this phase which is characteristic for that creature under the particular conditions of cultivation. This represents the maximal reproductive potential of the organism. Each cell upon division produces two equal daughter cells, these two cells can in time give rise to four cells, these four in an equal period of time can give rise to eight, eight to 16, 16 to 32, and so on. The time required for each doubling of the cell population during exponential growth is called the *generation time*.

What are the factors which affect the rate of growth? Of course, the primary factor is the organism itself. It takes *E. coli* about 20 minutes to double its mass under the best growth conditions. However, a rabbit at birth requires approximately 6 days to double its mass; a guinea pig at birth, 18 days; a human child at birth, 180 days.

Another factor of major significance is the medium. If *E. coli* is grown in a very rich medium containing all amino acids, purines, pyrimidines, vitamins, and other nutrients, it will have a generation time of about 20 minutes at 37°C. The same organism, however, when placed in minimal medium with glucose as the sole carbon and energy source, will have a generation time of 50 or 60 minutes. The reason for this is that in a rich medium the organism does not have to synthesize the myriad enzymes required for the biosynthesis of all of the amino acids, purines, vitamins, and so on. It uses preferentially those nutrients from the medium and thus it conserves energy and carbon. This conserved energy and carbon are used for growth and division rather than for the biosynthesis of building blocks for macromolecules. This phenomenon will be discussed further when we consider regulation of enzyme synthesis.

Temperature is another factor which greatly influences the rate of growth. The primary effect of temperature on cellular growth is that it affects the rate of enzyme reactions. Every enzyme has an optimum temperature for maximal activity. Reducing the temperature below the optimum reduces the rate of enzyme activity. Raising the temperature above the optimum results in reduced enzyme activity and eventually in destruction of the enzyme. Since growth is the result of the activities of enzymes working in harmony, an optimum temperature for growth is also observed. Any deviation from the optimum temperature results in a reduction in the rate of growth. Optimum temperatures for growth vary dramatically. For example, algae have been found growing in the Yellowstone Geysers where temperatures are almost at the boiling point. At the opposite extreme it is not at all uncommon to find bacteria and fungi growing, albeit very slowly, in icehouses and refrigerators.

Exponential growth cannot proceed indefinitely, either in nature or in the laboratory. The consequences of prolonged exponential growth would be disastrous. A single *E. coli* cell with a generation time of 20 minutes would produce in 48 hours of exponential growth 2.2×10^{43} cells! The total weight would be about 2.5×10^{25} tons, or roughly 4000 times the entire mass of the earth.

What are the factors that cause exponential growth to terminate and bring on the maximum stationary phase? The two most common factors are the exhaustion of nutrients and the production of toxic metabolic wastes by the culture. During the process of growth and division, nutrients are constantly being assimilated by the cells, hence the concentration of these nutrients in the medium is being depleted. When the concentration of the limiting nutrient is exhausted, growth ceases. Under these circumstances, then, the extent of growth will be directly proportional to the concentration of the limiting nutrient. In the laboratory exponential growth may terminate as a result of oxygen deprivation. In cultures of high population densities the rate of oxygen utilization may become greater than the rate of oxygen diffusion into the medium. As a result the culture may become starved for oxygen and exponential growth ends.

Also, during growth, organisms produce and excrete metabolic wastes which may reach concentrations that are toxic. It is an interesting general point that as a consequence of growth, the environment becomes less and less favorable for further growth to occur; the growth rate decreases and eventually cell division ceases, and the maximum stationary phase is established.

Exponential growth may also be terminated by artificial means. For example, the sulfa drugs are used to inhibit the growth of many pathogenic (disease producing) bacteria. As shown in Figure 7.2, the sulfa drugs resemble chemically para-aminobenzoic acid (PABA), which is an integral part of one of the B vitamins, folic acid. One of the enzymes needed for the synthesis of folic acid cannot distinguish between sulfa drugs and PABA. The enzyme combines with the drug and is therefore unavailable for the synthesis of the folic acid precursor. Thus if sulfa drugs are added to an exponentially growing culture of bacteria, growth soon ceases because folic acid is essential for the growth process. If folic acid is essential for growth of bacteria, and we have said previously that there is a unity in biochemistry, then why are not sulfa

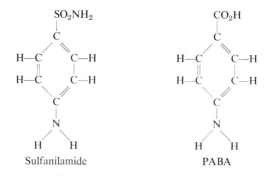

FIGURE 7.2 The chemical similarity between one of the sulfa drugs, sulfanil-amide, and para-aminobenzoic acid (PABA).

drugs toxic to man as well? The reason is that man requires folic acid pre-formed in his diet—man is totally deficient in the enzyme that is inhibited by the sulfa drugs. The use of chemical compounds which specifically inhibit the invading pathogen but have no deleterious effects on the host is the basic tenet of chemotherapy.

The stationary phase of growth

In this phase of the growth curve the population has reached the maximal level that the environment permits. This period may last for hours or even for days. Once the stationary phase has been attained, organisms are no longer involved with the problems of growth and division but are now faced with the more immediate problem of survival. Energy is required for this survival process—for the maintenance of proper osmotic pressures, for motility, for the resynthesis of proteins which are constantly being broken down during the stationary period, and so on. This is termed the energy of maintenance.

If growth has ceased due to the complete exhaustion of the carbon and energy source in the medium, where does the cell obtain energy for maintenance? Most cells accumulate storage products during exponential growth. In liver cells, for example, this storage product is in the form of glycogen, in plant cells it is starch, and in microbial cells either glycogen, starch, or other organic storage materials. Storage products are of large molecular size and are usually stored in the form of granules. Once cells reach the stationary phase, these storage materials are degraded to obtain the essential energy for maintenance. When these storage products have been consumed, the cell is faced with the necessity of degrading other cellular components to obtain energy for survival. It is apparent that this destruction of cellular constituents cannot go on for long without having deleterious effects. Death will soon ensue.

The death phase

In this period of the growth curve, the numbers of viable cells, that is, the cells capable of giving rise to progeny, decrease sharply. Populations contain individuals (presumably mutants) having varying degrees of resistance. Those creatures most susceptible to the adverse environment will, of course, die first; the more resistant ones will have a prolonged viability. It is often observed that after the majority of the population has died, the rate of death decreases, and a small number of survivors may persist for many months or even years. This may be due to cannibalism. Those cells which have survived grow and divide at the expense of the nutrients released from the decomposition of the dead cells.

Some microbial cells produce resistant forms known as *spores*. One of the outstanding features of spores is their tremendous resistance to adverse environmental conditions. Viable spores, for example, have been isolated from mummies buried in Egypt over 2000 years ago.

7·3 Diauxic growth

In 1941 Jacques Monod of the Pasteur Institute in Paris published a monograph dealing with studies on bacterial growth. The last sections of this work describe experiments in which growth was measured in media containing mixtures of two sugars. Monod observed that although growth was normal in the presence of certain sugar mixtures, that is, a single exponential phase was seen as illustrated in Figure 7.3, two distinct exponential phases separated by a short lag were elicited by other sugar mixtures (Figure 7.4). He named this phenomenon *diauxic growth* (double growth).

Monod conducted numerous experiments in which the relative concentrations of two sugars eliciting diauxic growth were varied. Typical data obtained for the mixture of glucose and sorbitol are presented in Figure 7.5. The extent of growth during the first cycle was directly proportional to the concentration of glucose; essentially three times as much growth was observed in the first phase when the glucose concentration was tripled. Similarly, the extent of growth in the second growth phase was proportional to the concentration of sorbitol. These experiments suggested that the initial growth phase represented growth as a result of the exclusive utilization of all of the glucose in the medium prior to initiation of growth at the expense of sorbitol. Proof of this hypothesis was obtained when chemical analyses revealed that all of the glucose had been consumed during the first growth cycle. It appeared, then, that bacteria could discriminate between two closely related compounds,

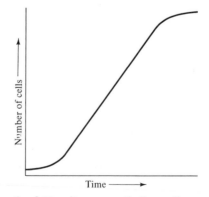

FIGURE 7.3 The growth of *E. coli* on a synthetic medium containing a mixture of the sugars glucose and fructose.

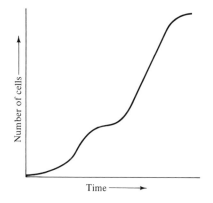

FIGURE 7.4 The growth of *E. coli* on a synthetic medium containing a mixture of the two sugars glucose and lactose.

glucose and sorbitol, and that they would utilize preferentially glucose prior to utilization of sorbitol. After extensive experimentation, it was learned that the utilization of sorbitol took place only after an induction period during which the enzymes responsible for the metabolism of this sugar had been produced by the cell.

These pioneering experiments by Monod provided the impetus for research on the regulation of enzyme synthesis for which he later (1965) shared the Nobel prize with two of his colleagues at the Pasteur Institute, François Jacob and André Lwoff.

Before Monod's discoveries, other investigators had observed temporary changes in enzymatic activities of microorganisms in response to changes in the growth medium. H. Karstrom in 1930 formulated the concept that bacterial cells manufactured two types of enzymes in response to environmental changes: *constitutive enzymes* were those produced by the cell at all times, independent of the nature of the medium, whereas *inducible enzymes* (called adaptive by Karstrom) were elicited by cells in response to the presence of a specific chemical in the environment. A chemical which induces the production of a specific enzyme by a cell is now called an *inducer*.

It should be emphasized that the appearance of inducible enzymes was not a result of mutation and selection of a new cell type; rather it was a temporary change in the expression of existing genetic potential. That is, no new genetic information had been introduced into the population; a change in the environment induced the expression of preexisting genes. This implies that cells are able to regulate, within limits, their enzymatic constitution in response to changes in the environment. The number of inducible enzymes that a cell can produce, of course, is dictated by its genetic potential, that is, a cell can elicit an inducible enzyme only if it has the genetic information requisite for the structure of the specific enzyme-protein.

The capacity to alter its enzymatic constitution in response to environmental changes is of profound significance to the survival of cells. Microbial cells are in direct contact with their environment which may undergo drastic fluctuation in composition. Let us consider, for example, an *E. coli* cell growing on sorbitol. The culture is transferred to a medium which contains

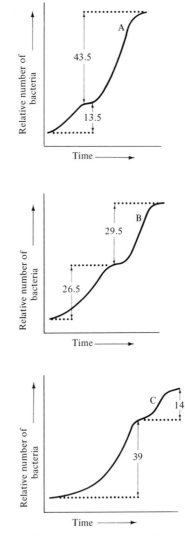

FIGURE 7.5 The growth of *E. coli* on a synthetic medium containing a mixture of the two sugars glucose and sorbitol. (a) 1 part glucose: 3 parts sorbitol; (b) 2 parts glucose: 2 parts sorbitol; (c) 3 parts glucose: 1 part sorbitol. The numbers represent the relative growth yield.

the sugar lactose instead of sorbitol. The cells respond to this change in nutrients by inducing the synthesis of the enzymes responsible for metabolizing lactose. Since sorbitol is no longer present in the medium, the enzymes responsible for its metabolism are no longer synthesized. Induced enzyme formation allows cells to produce many enzymes, but *only* when their function is required for survival.

7•4 Induced enzyme formation

Induction may be defined as the increased rate of enzyme-protein synthesis under the influence of specific substances (inducers) unaccompanied by genotypic changes. The induced enzyme which has been most extensively studied is the β-galactosidase of *E. coli*. The enzyme functions by splitting the sugar lactose into its constituent monomers, glucose and galactose. An experiment to demonstrate the induction of β-galactosidase is performed as follows: Inducer is added to a culture of *E. coli* growing exponentially on a minimal medium (Figure 7.6, lower arrow). The increase in β-galactosidase activity is measured as a function of growth. At the time marked by the upper arrow the inducer is removed.

Several important conclusions may be deduced from this experiment: (1) before the addition of inducer the level of enzyme in the culture is very low (basal level); (2) shortly after addition of the inducer the amount of enzyme activity increases dramatically and continues to rise at a constant rate of as much as 1000 times the uninduced value as long as the inducer is present; (3) after removal of the inducer no further increase in enzyme activity is observed—the enzyme level remains constant. This type of experiment

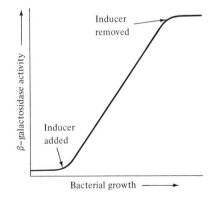

FIGURE 7.6 Induced synthesis of β-galactosidase by *E. coli*. Inducer is added (lower arrow) to an exponentially growing culture of *E. coli*. β-galactosidase activity is measured in samples of the culture. Later inducer is removed (upper arrow).

describes in general terms the induction process, but provides little information about the mechanism, that is, the precise molecular basis of induction.

Regulation of β-galactosidase activity in response to the inducer might occur at one of three possible steps in the synthesis of proteins: at the transcriptional level (synthesis of mRNA); at the translational level (synthesis of polypeptide chains); or possibly at the activation or conversion of an inactive protein to an active β-galactosidase.

The precursor for β-galactosidase. Many attempts were made to resolve the question of the site of regulation of β-galactosidase synthesis. The following data provided circumstantial evidence that the regulation was not at the activation of an inactive protein.

1. Energy was required after addition of inducer before β-galactosidase activity appeared. For example, in the presence of certain compounds such as sodium azide, cyanide, and dinitrophenol, which inhibit ATP generation, β-galactosidase was not formed.
2. Cells starved of amino acids failed to be induced for β-galactosidase.
3. The drug chloramphenicol, which specifically inhibits protein synthesis, also inhibits the induction of the enzyme.

Since energy, as well as amino acids, was required for enzyme induction, and since chloramphenicol also prevented β-galactosidase synthesis, these experiments suggested but did not prove that enzyme synthesis was a result of *de novo* protein synthesis (synthesis of protein from their component amino acids). In 1955 this question was answered conclusively by decisive experiments carried out independently at the Pasteur Institute and at the University of Illinois. *Escherichia coli* was grown in a minimal medium in the presence of radioactive sulfur (^{35}S) but in the absence of inducer. As a result of growth in the presence of ^{35}S, cellular proteins became radioactively labeled. After removal of all the ^{35}S which was not incorporated into proteins, the culture was transferred to a medium containing inducer and nonradioactive sulfur (^{32}S), and growth was permitted to continue. The β-galactosidase was isolated and purified and the level of radioactivity in the protein was determined. The β-galactosidase contained insignificant levels of radioactivity. If the enzyme had been synthesized from preexisting protein precursors, the purified β-galactosidase would have been radioactively labeled, because the precursors were labeled with ^{35}S during the initial part of the experiment. Since the enzyme was not found to contain radioactive ^{35}S, this implied that the β-galactosidase was synthesized exclusively from materials assimilated after the addition of inducer. Hence proteins present in the noninduced cells did not serve as precursors of the active enzyme. In summary, induced enzymes are synthesized completely *de novo* from their respective amino acids; control of induced enzyme synthesis must therefore reside at the transcription or translational level.

Galactoside permease

Investigations carried out by Monod and his collaborators provided a clue to the site of regulation. They tested a large number of compounds related to lactose as inducers and as substrates (compounds split by the enzyme) for β-galactosidase. Certain compounds behaved both as inducers and substrates for the enzyme; other compounds were strictly inducers and did not act as substrates; still others behaved as substrates but were incapable of inducing the enzyme. Those molecules which acted as inducers but which were not metabolized by the enzyme were called *gratuitous inducers*. The use of gratuitous inducers eliminated the problems associated with metabolizable substrates. Since gratuitous inducers are not metabolized, their concentration can be specified precisely throughout the experiment. In addition, by using gratuitous inducers, the Paris group discovered a new protein involved in the metabolism of lactose by *E. coli*. When radioactively labeled gratuitous inducers were added to a fully induced culture, there was a rapid accumulation of the labeled inducer in the cells. Bacteria not previously induced did not accumulate the compound. These, and many genetic experiments. led them to conclude that the accumulation of inducer was caused by the action of an enzyme protein, distinct from β-galactosidase, which they called *galactoside permease*. Permeases, in general, function in the transport and concentration of specific compounds from the medium into the cell. Galactoside permease was found to be inducible. Furthermore, its synthesis was controlled by the same inducer as β-galactosidase.

A third protein, *galactoside transacetylase*, was later discovered in Paris. This protein was also found to be inducible, and its activity appeared almost simultaneously with the activities for β-galactosidase and permease. Thus three independent and distinct proteins are synthesized when a single inducer is added to a culture of *E. coli*.

Genetics

Investigations employing the refined tools of modern biochemical genetics furnished another major clue to the mechanism of induction.

The Paris group under the leadership of Jacob and Monod isolated a large number of mutants of *E. coli* that were deficient in some aspect of lactose metabolism. Characterization of these mutants revealed the existence of *three distinct genes* involved in lactose utilization: the z gene determines the structure of the β-galactosidase protein; the y gene determines the structure of β-galactoside permease; and the a gene the structure of transacetylase. Mutants with all possible combinations of genotypes were obtained. For example, $z^+y^+a^+$ cells possess the genetic information for the synthesis of all three enzyme proteins upon induction (β-galactosidase, permease, and transacetylase); $z^-y^+a^+$ have the genetic information for making permease and transacetylase but not β-galactosidase.

In addition to mutants lacking the potential for producing one or more of the enzymes, they isolated another class of mutants called *constitutives* for β-galactosidase, that is, the enzyme was synthesized in the *absence* of inducer. Surprisingly, these constitutive mutants synthesized not only β-galactosidase in inducer-free medium but also permease and acetylase, even though the structure of these three different proteins was specified by three distinct genes. This observation led to the suggestion that the mutation to consti- tutivity had taken place in still another gene (*i*) distinct from the other three (*z*, *y*, and *a*). Mutation of the *i* gene from i^+ to i^- allowed the *expression* of latent z^+, y^+, and a^+ genes in the absence of inducer. This was a striking phenomenon without precedence in biology—one genetic element regulating the expression of three others.

All bacterial genes described previous to this time functioned in deter- mining the amino acid sequence of a protein (structural genes). Since proteins are relatively easy to isolate and characterize, structural genes are therefore amenable to study. By investigating the alteration in the protein one could study the end result of a mutation. But how does one study a regulatory gene such as the *i* gene? The mechanism of bacterial conjugation which had just been clarified by Jacob and Wollman in Paris provided an experimental tool for these studies. It was known that in bacterial conjugation only the DNA of the male ($\male$) bacterium was injected into the female ($\female$) recipient without any cytoplasmic exchange. The zygote that is formed thus contains genes from both the $\male$ and $\female$, but the cytoplasm is entirely female. In 1959 Arthur Pardee, along with Jacob and Monod, performed an ingenious experiment designed to investigate the expression of the i^+ and z^+ genes when injected into a female bearing the z^-i^- alleles. Basically, the question was whether the i^+ gene product could exert some controlling effect on the expression of the *z* gene. This famous experiment is affectionately called the PaJaMo experi- ment in reference to the authors. The formation of β-galactosidase by zygotes was measured. Several bacterial crosses were performed:

A. $\male$ (z^+i^+) × $\female$ (z^-i^-)

Neither male nor female could synthesize the enzyme in the *absence of inducer*: the $\male$ because it was i^+ (inducible); the $\female$ because it was z^- (that is, it lacked the genetic information for making the enzyme). The zygotes (z^+i^+/z^-i^-), however, did synthesize the enzyme immediately after DNA injec- tion in the absence of inducer (constitutive synthesis) at a rate similar to the synthesis in the presence of inducer. The reciprocal mating was then per- formed and the zygotes similarly tested.

B. $\male$ (z^-i^-) × $\female$ (z^+i^+)

For the same reasons mentioned neither parent synthesized the enzyme in the absence of inducer. In this cross, however, no trace of β-galactosidase was detected at any time after mating in the *absence of inducer*. The zygotes formed in matings A and B were genetically identical (z^+i^+/z^-i^-), yet zygotes from mating A produced enzyme constitutively, whereas zygotes from mating B did not! The only difference furnished by the ♀ cell, was i^- in cross A and i^+ in cross B. These striking results suggested that i^+ gene produced a cytoplasmic substance which inhibited the expression of the z^+ gene (mating B), whereas in mating A the ♀ cytoplasm was i^- and thus did not produce this gene product. This interpretation suggested that the i^+ inducible allele was dominant and the i^- constitutive allele was recessive. This predicts that an i^- ♀ cytoplasm, lacking the cytoplasmic substance, should become inducible after receiving the i^+ gene from the ♂, but only after the i^+ gene has had an opportunity to be expressed (that is, transcribed and translated in the zygote). To test this hypothesis cross C was performed.

C.   ♂ $(z^+i^+) \times$ ♀ (z^-i^-)

In this experiment, after zygote formation, one-half of the zygotes received inducer and the other half did not. The male donors were killed off shortly after conjugation so that upon addition of inducer only the zygotes would make enzyme. The results shown in Figure 7.7 demonstrate that in the absence of inducer the zygotes synthesized β-galactosidase constitutively for about 2 hours after mating. At that time enzyme synthesis stopped unless inducer was added. This confirms the prediction that the i^+ gene exerts its

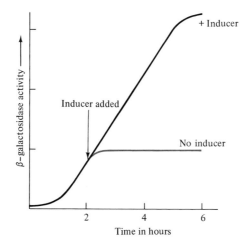

FIGURE 7.7 β-galactosidase production in zygotes of *E. coli* formed by mating ♂ $(z^+\ i^+) \times$ ♀ $(z^-\ i^-)$.

effect in the female recipient by producing a substance which is responsible for rendering i^+z^+ cells inducible.

In summary, the i^+ gene directs the synthesis of a cytoplasmic substance which prevents the production of β-galactosidase (permease and trans-acetylase) unless inducer is added to the medium. In i^- mutants no such substance is produced (or if produced, it is not functional), and the β-galactosidase (permease and transacetylase) are formed even in the absence of inducer. Pardee, Jacob, and Monod proposed that (1) the i^+ gene product was a specific *repressor* which inhibited the synthesis of β-galactosidase; (2) inducers counteract the action of the repressor, permitting enzyme formation. Since constitutive mutants did not make the repressor, inducer was not necessary for enzyme formation.

The operon model

As a consequence of these experiments, Jacob and Monod in 1961 proposed the *operon model* for the regulation of gene expression (Figure 7.8). According to this model, the information from the three structural genes (z^+, y^+, a^+) is transcribed into mRNA. Translation of this information then takes place on the ribosome, resulting in the three enzyme protein products. The i^+ gene exerts its effect on this system by directing the synthesis of a repressor which acts on the DNA at a site called the *lac*tose operator (o) region. The reaction between repressor and lac operator region prevents transcription. Inducers somehow prevent the repressor from combining with the operator region, thus permitting mRNA synthesis. In essence, reaction of the repressor with the operator prevents transcription. Inducers function by interfering with the reaction between repressor and operator, thus allowing transcription to occur. The three structural genes whose expression is regulated by an operator plus the o region were called the lac operon, thus the name, the operon model.

The operon model makes several predictions. It predicts that (1) the repressor is the product of the i^+ gene; (2) the repressor can combine with the operator region (o) of the lac operon; (3) the repressor must also react with the inducer (for mRNA synthesis to take place the inducer must be able to displace that repressor which is in combination with the operator region); (4) when the repressor is in combination with the operator, no mRNA synthesis for enzymes of the lactose operon will take place; (5) when in combination with the inducer, the repressor is inactive in interfering with the synthesis of lac operon mRNA and thus enzyme formation can take place.

Before the operon model could be experimentally verified, the repressor substance had to be isolated. In December 1966 Walter Gilbert and Benno Müller-Hill of Harvard University reported the isolation of the lac repressor.

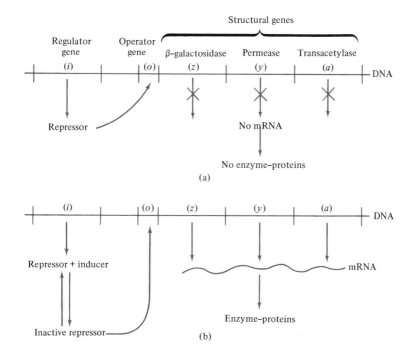

FIGURE 7.8 The operon model. (a) The *i* gene product (repressor) combines with the *o* region and prevents mRNA synthesis. (b) Inducer reacts with the repressor and inhibits the reaction of repressor with the *o* region. mRNA synthesis can thus take place.

The repressor was found to be a protein. Verification of these predictions soon followed.

Evidence that the i^+ gene product was the repressor was obtained using i^- mutants that either manufactured faulty repressor or failed to produce repressor completely. Repressor could be isolated from i^+ strains but not from the i^- mutants. The first prediction was thus verified—the i^+ gene product is the repressor.

DNA from cells having the lac operon (lac$^+$) was isolated and purified. The lac repressor protein was shown to combine with this lac$^+$ DNA. DNA from cells which did not possess the lac operon (lac$^-$) did not bind the lac repressor. Furthermore, DNA isolated from cells having a mutation at the operator region of the lac operon failed to bind the repressor. Thus it was shown that the repressor not only combined with DNA but it appeared to combine specifically with the operator region of the lac operon.

The initial isolation of the *lac* repressor was based on the prediction that it had to react with inducers of the lac operon. Gilbert and Müller-Hill isolated a protein from an inducible strain of *E. coli* which combined with radioactive inducer. Moreover, they showed that adding inducer to a repressor-DNA complex released the repressor from the DNA.

In 1968 a research group at the University of Wisconsin reported that isolated repressor inhibited mRNA synthesis in vitro. Presumably, inhibition of mRNA synthesis by the isolated repressor took place by blocking DNA transcription.

Several independent lines of evidence have demonstrated that addition of inducer to an i^+ cell resulted in the synthesis of mRNA for the lac operon. This mRNA was not made unless the inducer was added.

It is apparent that all the predictions derived from the operon model have been experimentally verified. Although this discussion has been restricted specifically to β-galactosidase and the lac operon, investigations on the mechanism of induction of other enzymes reveal that the operon model may well be a general mechanism for enzyme regulation. The concept that regulatory genes, such as the *i* gene, function by directing the synthesis of repressors which regulate DNA transcription in response to environmental changes is probably the underlying mechanism of enzyme regulation.

7·5   Repression

A regulatory mechanism of even greater economic significance to the cell than induction is that of *repression*—the effective *inhibition of formation of enzymes* of a biosynthetic pathway by the endproduct of that pathway. For example, *E. coli* when growing in a minimal medium with glucose as the carbon and energy source must produce all of the enzymes involved in the biosynthesis of the amino acid isoleucine (the endproduct of the biosynthetic pathway). However, if isoleucine is added to the growth medium, *none* of the enzymes responsible for the synthesis of this amino acid is produced. The formation of enzymes of other biosynthetic pathways is not affected. In other words, the presence of isoleucine in the medium specifically represses the formation of its own biosynthetic enzymes.

The phenomenon of repression has been observed in numerous biosynthetic pathways in bacteria. Regulation of the enzymes of the histidine biosynthetic pathway has been extensively investigated. Histidine is an amino acid found in many proteins; it is therefore essential to the cell. The biosynthesis of histidine requires eleven enzymatically catalyzed reactions. Because two of the enzymes each carry out two steps in the pathway, nine structural genes are involved in the formation of these enzymes. These biosynthetic enzymes function in harmony; the product of one enzyme serves as the substrate for the next enzyme in the pathway, and so on, until histidine is formed. Histidine is then used for the biosynthesis of proteins.

Structural genes for enzymes of several biosynthetic pathways have been found in clusters on the chromosome; each cluster has its own operator region. The genes determining the structure of the enzymes in the histidine pathway are also clustered on the chromosome of some bacteria, that is, they lie adjacent to each other on the DNA. All nine histidine structural genes are under the control of one operator region. Thus we have a *his*tidine operon similar to the lac operon. The various genes in the operon are transcribed in a coordinated fashion. For example, bacteria make slightly more histidine than is required for protein synthesis. If, however, the intracellular concentration of histidine becomes limiting, all of the enzymes in the histidine biosynthetic pathway are produced in increased and equal quantities (*coordinate derepression*). The ultimate effect is a harmonious and coordinated increase in all of the enzymes of the biosynthetic pathway, resulting in an increased rate of histidine production to meet the increased demand. When histidine is no longer required, either because of reduced demands or because it is supplied to the culture, the formation of *all* of the enzymes in the pathway is repressed coordinately (*coordinate repression*). The cell makes use preferentially of nutrients supplied to it rather than expend the energy and the carbon required to make them. It is unnecessary for the cell to manufacture the numerous enzymes in the biosynthetic pathway for this endproduct, and it can channel this conserved energy for other uses. Thus regulation of enzyme synthesis by repressing the formation of biosynthetic enzymes ensures the cell of balanced amounts of precursor molecules for macromolecular synthesis.

The molecular mechanism of repression has been conceived as a modification of the mechanism described for induction (Figure 7.9). A repressor substance (most probably a protein) is produced by the regulatory gene of the biosynthetic operon (similar to the *i* gene of the lac operon). In its native state this repressor substance is inactive. The inactive repressor has no effect on the transcription of the structural genes of the operon since it cannot combine with the operator region of the operon. However, when the endproduct of the biosynthetic pathway is present in excess, it combines with the inactive repressor and changes it to an active repressor. The active repressor can then combine with the operator region and prevent transcription of the structural genes of the operon.

Repressors, then, may exist in either active or inactive forms. In inducible systems, such as the lac operon, combination of the repressor with the inducer rendered the repressor inactive and nonfunctional; in repressible systems, such as the histidine operon, combination of the inactive repressor with histidine results in an active, functional repressor.

7·6 Feedback inhibition

We have discussed cellular mechanisms of regulation which function by turning on or turning off the transcription process. These mechanisms permit the formation of enzymes when they are required (induction) or prevent

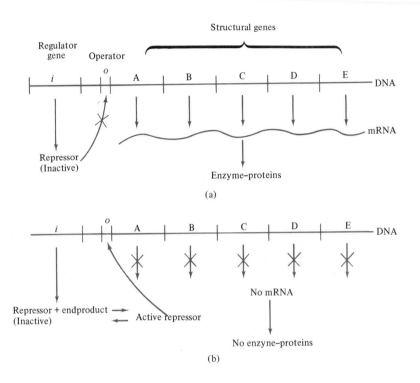

FIGURE 7.9 Current view of repression of enzyme formation. (a) The inactive repressor protein product of the *i* gene cannot combine with the *o* region; mRNA is formed and translation occurs. (b) The inactive repressor reacts with the end-product and is converted to an active repressor which can combine with the *o* region, thus blocking mRNA synthesis.

enzyme formation when they are no longer necessary (repression). A third regulatory mechanism which has been extensively studied is called *negative feedback inhibition*. This control device functions by an *inhibition of the activity* (not the formation) of the first enzyme in a biosynthetic pathway by the endproduct of that pathway. Thus it is complementary to the repression of enzyme formation previously discussed (Figure 7.10).

One of the earliest examples of negative feedback inhibition was reported in 1953 by H. Edwin Umbarger working at the Long Island Biological Laboratories in New York. Umbarger found that when *E. coli* was grown on a minimal medium with glucose as the carbon and energy source, the organism produced the amino acid, isoleucine. If, however, excess isoleucine

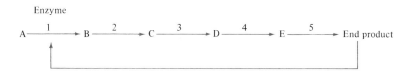

FIGURE 7.10 Feedback inhibition of a biosynthetic pathway by the endproduct of that pathway. The endproduct prevents the conversion of A to B by inhibiting the activity of enzyme 1.

was fed to the culture, the bacteria preferentially used the supplied amino acid and *immediately* ceased to produce their own.

Feedback inhibition is a remarkably efficient and economical control device for the cell. The *instant* that the cellular concentration of isoleucine reaches an adequate level, the cell *immediately* stops making more of it. The specificity of the system became apparent when Umbarger demonstrated that the first enzyme in the biosynthetic pathway, and *only* the first, was inhibited by the endproduct. Since, as pointed out previously, biosynthetic enzymes function in tandem (one enzyme forming a product which becomes the substrate of the next enzyme in the pathway, and so on) inhibition of the first enzyme effectively shuts off the activities of all the enzymes in the sequence. Moreover, the inhibition of the first enzyme in the pathway by the endproduct need not be absolute. Rather, it may be a partial inhibition depending on the relative level of endproduct present (Figure 7.11). This fine control of endproduct formation is crucial for the harmonious functioning of the cellular machinery, since the rate of isoleucine utilization (or

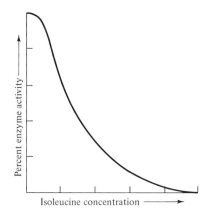

FIGURE 7.11 Inhibition of the first enzyme in the isoleucine biosynthetic pathway by varying concentrations of isoleucine. As the concentration of isoleucine increases, progressively greater inhibition is observed.

other endproducts) for protein synthesis may vary from time to time. Only the endproduct of the pathway exerts the inhibitory effect on the first enzyme of the pathway. Even compounds very closely related to isoleucine have no effect.

The observation that the endproduct of the biosynthetic pathway can inhibit not only enzyme activity but also enzyme formation might indicate that repression and feedback inhibition are somehow interrelated. Curiously, the two regulatory mechanisms are completely independent. Mutants of *E. coli* have been isolated which are no longer susceptible to feedback inhibition of the first enzyme in the isoleucine pathway, but which remain susceptible to repression of the isoleucine biosynthetic enzymes. Conversely, mutants which are no longer repressible remain susceptible to feedback inhibition. In addition, the two mutations occurred at different sites on the chromosome. Therefore feedback inhibition and repression of the same biosynthetic pathway are independent but complementary regulatory mechanisms.

Feedback inhibition has been demonstrated for many different biosynthetic pathways; several amino acids, vitamins, and the purine and pyrimidine biosynthetic pathways have been shown to be regulated by feedback inhibition. It appears to be a general mechanism for regulating the concentration of intracellular building blocks.

In feedback inhibition the endproduct inhibits the activity of its first biosynthetic enzyme. This is a negative control. Numerous examples of *positive control mechanisms* have also been reported, that is, the regulatory mechanism *activates* (instead of inhibiting) an enzyme when the conditions in the cell warrant it. An instance of such a positive control mechanism is the regulation of the level of ATP in cells. Exponentially growing cells use some of the nutrients (for example, glucose) supplied to them for carbon and energy. A portion of the supplied foodstuffs is stored by the cells in the form of glycogen or starch, and so on. Under growing conditions the cellular level of ATP is maintained in a careful balance. The amount of energy in the form of ATP which is expended by the cell (for synthesis, and so on) is replenished by the burning of foodstuffs by way of respiration. When the concentration of ATP in the cell is reduced to a low level (possibly due to exhaustion of the supplied nutrients), the enzyme responsible for degrading the stored glycogen is activated. The glucose obtained from the breakdown of glycogen is then used for respiration and ATP generation. What is the molecular signal that stimulates glycogen breakdown? When ATP is used for energy-requiring functions, ADP (adenosine diphosphate) is formed along with some AMP (adenosine monophosphate). AMP acts as the signal which activates the glycogen-degrading enzyme so that the ATP level may be replenished. During the process of ATP synthesis, AMP levels are reduced since AMP is a precursor of ATP. Thus the AMP levels are lowered, and the signal to stimulate glycogen breakdown is turned off.

7·7 Allosteric interactions: a unifying concept

The molecular mechanism of regulation of enzyme activities, either by negative control (inhibition) or positive control (activation), has been investigated in several laboratories in recent years. It has long been known that enzymes participate in biological reactions by combining with their specific substrates. A specific site on the enzyme, called the *active site*, is responsible for this union with substrate. In 1962 John Gerhart and Arthur Pardee at the University of California were investigating the feedback inhibition of the first enzyme in the pathway of pyrimidine biosynthesis. They discovered that the isolated and purified enzyme remained susceptible to inhibition by the endproduct of the pathway. Gerhart and Pardee learned that their purified enzyme not only had an active site for combining with its substrate but also had a different site for combining with the feedback inhibitor. They demonstrated that these two combining sites were quite distinct; abolishing the combining site for the inhibitor had no effect on the active site for the substrate. The two sites, although on separate parts of the enzyme, could exert their effects on each other; that is, they were functionally linked. To explain this phenomenon, Monod, Jean-Pierre Changeux, and Jacob proposed the hypothesis of *allosteric interactions*. They suggested that a class of proteins (*allosteric proteins*) existed whose three-dimensional structure was altered when combined with certain compounds (inhibitors or activators). In this altered structural form, the enzymatic activity of the proteins was either inhibited or activated. This combination was called *allosteric interaction*. For example, during endproduct inhibition of an enzyme, it is presumed that the inhibitor combines with the enzyme at its specific site and changes the appearance of the enzyme. The active site for the substrate is then no longer a perfect "fit" (Figure 7.12a). This alteration in the active site would reduce the activity of the enzyme since it could no longer bind to the substrate at a normal rate. Conversely, in the case of positive control, where enzymes are activated, the combination of a small molecule (AMP) with the enzyme would change the appearance of the enzyme so that the "fitness" for the substrate is improved (Figure 7.12b).

Allosteric interactions may also be invoked to explain the activation or inactivation of repressors by inducers or endproducts. The inducer would then act by altering the three-dimensional appearance of the repressor so that the combining site for the operator would be changed. The endproduct would, on combining with an inactive repressor, improve the "fitness" of the combining site for the operator region.

It should be evident from the foregoing discussion that cells have developed a variety of regulatory mechanisms for coping with changes in their environment. The complex processes of growth and reproduction require highly integrated systems for the production of balanced levels of

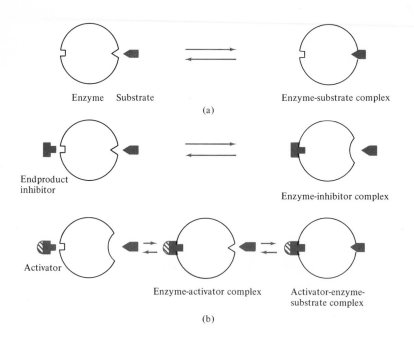

FIGURE 7.12 Schematic representation of allosteric interaction. (a) Allosteric enzyme and endpoint inhibitor. (b) Allosteric enzyme and activator. In (a) the feedback inhibitor reacts with the enzyme and alters the "fitness" of the active site so that the substrate and enzyme can no longer combine. In (b) the activator improves the "fitness" of the active site for the substrate by combining with the enzyme.

building blocks for the synthesis of cellular macromolecules. An imbalance of these levels is obviously detrimental. Underproduction of a precursor results in a reduction of macromolecular synthesis, which would result in a reduction in the growth rate; overproduction of a precursor is wasteful since the cell cannot accommodate high concentrations of building blocks and thus excretes them.

The integrated operation of repression and feedback inhibition furnishes the cell with a mechanism for adjusting the levels of building blocks in relation to cellular needs and compensates for alterations in the environment. By reducing the levels of biosynthetic enzymes via repression, overproduction of building blocks would not be prevented immediately. The reason for this is that the existing enzymes would continue to function and the endproduct would continue to be made even though it was in excess or no longer needed. This would be wasteful synthesis, a crime that cells attempt not to commit. Endproduct inhibition, on the other hand, would operate immediately, to stop cellular synthesis of the building blocks when they are present in adequate supply.

The generalized control of cellular metabolism by allosteric interactions, then, provides the cell a mechanism for adjusting the expression of its genetic potential to cope with variations in its environment. An allosteric change in the repressor enables the cell to invoke the synthesis of inducible enzymes when their substrate suddenly appears in the medium. When that substrate has been consumed, an allosteric change in the repressor then shuts off further synthesis of these no longer needed enzymes. At the end of the exponential growth period, terminated possibly because of nutritional inadequacy in the environment, the endproducts of biosynthetic enzymes would accumulate unless an allosteric interaction with the inactive repressor inhibited the further synthesis of the enzymes. Moreover, it is by still another allosteric interaction with the first enzyme in the biosynthetic pathway that the endproduct immediately turns off its further synthesis. *It is clear therefore that the relative composition of the environment acts as a switch which triggers the expression of the genetic information of the cell by allosteric interactions.*

The problem of regulation in multicellular organisms is more complex than in bacteria. We know, for example, that in a human being there are liver cells, brain cells, kidney cells, nerve cells, all looking and behaving very differently. Yet to the best of our knowledge, all these cells have identical copies of DNA. Although the (potential) information is the same, the cells turn out to be different. It follows that there must be some sort of switch, something which turns on and off particular genes in appropriate environments. The process by which a single fertilized egg divides into two, then four, then eight, and so on, until an aggregate containing different cell types is formed is called *differentiation*. At present the underlying control mechanisms for this remarkable process is unknown.

As we have discussed in this chapter, even single cell organism, such as *E. coli*, have elaborate control mechanisms. Great variation can exist between the amount of a specific enzyme present when it is needed and when the environmental conditions are such that it would serve no useful purpose. We are just beginning to understand in a few simple cases how control systems operate. It will be interesting to see if information obtained on bacterial regulation will provide clues to the more complex problem of differentiation.

It is the nature of science that for each question answered several new and more fundamental problems arise.

As the area of light increases so does the circumference of darkness.

A. EINSTEIN

QUESTIONS AND PROBLEMS

7.1 A fresh medium is inoculated with 100 *E. coli* cells. Assuming that growth is logarithmic and that the generation time is 20 minutes, how long will it be until 1 million cells are present in the culture? How many cell divisions does this represent?

7.2 Why are there so many different types of bacteria, that is, why is not there a survival of the "fittest"?

7.3 The following enzyme system is inducible:

$$X \xrightarrow{\text{Enzyme 1}} Y \xrightarrow{\text{Enzyme 2}} Z$$

Enzymes 1 and 2 are specified by the structural genes *a* and *b*, respectively. Together with their regular gene (*i*) and operator gene (*o*), they occupy the following positions on the chromsome:

$$i \ o \ a \ b$$

State which enzymes will be present, with and without inducer, for the following zygotes:

$$\frac{i^+o^+a^-b^+}{i^+o^+a^-b^+}, \quad \frac{i^+o^-a^-b^+}{i^+o^+a^+b^-}, \quad \frac{i^-o^+a^+b^+}{i^+o^-a^-b^-}, \quad \text{and} \quad \frac{i^-o^+a^-b^-}{i^+o^-a^+b^+}$$

7.4 In 1969 man landed on the moon and brought back samples from its surface. Assuming that soil analyses revealed the total absence of any organic compounds and combined forms of nitrogen, what type of earthly organism would you expect to grow initially in the soil if it was not kept sterile? What two obligatory reactions would be prerequisite for these organisms to appear?

7.5 A logarithmically growing culture, having a doubling time of 2 hours and initially containing 4×10^8 cells per milliliter, uses 4 grams of glucose in growing 4 hours. How many grams of glucose is used in the first 2 hours? (It may be useful in solving this problem to make a rough graph of the growth of the culture over the 4 hours.)

SUGGESTED READINGS

Changeux, J.-P., "The Control of Biochemical Reactions," *Scientific American*, 212, 36 (1965). A discussion of repression with a more extended discussion of allosteric interactions.

Cohen, G. N., and J. Monod, "Bacterial Permeases," *Bacteriological Reviews*, 21, 169 (1957). An extensive discussion of bacterial permeases, lucidly written (in English).

Monod, J., "From Enzymatic Adaptations to Allosteric Transitions," *Science*, 154, 475 (1966). The Nobel lecture delivered in Stockholm December 11, 1965, when the author received the Nobel Prize in Physiology.

Umbarger, H. E., "Intracellular Regulatory Mechanisms," *Science*, 145, 674 (1964). This is an excellent account of feedback inhibition as described by the discoverer.

Watson, J. D., *The Molecular Biology of the Gene*, New York: W. A. Benjamin, Inc., 1965. A general treatment of molecular biology.

CHAPTER THE LAST

Which Is a Dialogue between Two Opposing Points of View, with Apologies to Socrates, and a Warning That All Characters Are Purely Fictitious, and That Any Resemblances to Scholars Living or Dead Is Purely Coincidental*

> I believe the intellectual life of the whole of Western Society is being split into two polar groups.
>
> C. P. SNOW

When the television screen finally cleared, the two men could see that the whole city had been obliterated by the strange blast. As the cameras swept over the land, they watched forest and countryside being engulfed by flames that were slowly but inevitably moving toward the secret deposit of n-tron munitions hidden in the middle of the tall trees. It was one of four hundred stockpiles gathered by Eutronia in event of attack from the Aggressians, who themselves had large stores of n-tron munitions. In a little while earth would erupt in a chain of inconceivable explosions. And there would be nothing but space, silence, and light where earth had been moments before. They had to escape.

The taller of the two men spoke. "My name is Thymine. Who are you?"

"Daedelus. I am poet, artist, seer, philosopher, believer, lover of the human. I sought the protection of this vault at the first sound of the strange roar."

"How appropriate for you to be named after the maker of the labyrinth of ages long past." And Thymine shuddered in memory of those ages. "But no matter. Take this suit and get into it quickly. We must leave or be destroyed."

Daedelus hesitated, but the look of fear in the other man's eyes made him obey. They slipped into the shiny blue polypropylene suits.

* This chapter was written in collaboration with John McMammon, Department of English, University of California, Los Angeles, California.

"Come on, man, hurry!" Thymine pulled him by the arm through a tunnel into a large green underground garden where a rocket was poised for a space thrust. "That's our only chance."

It was occurring to Daedelus that they were the last of the race. Time only to save themselves. Incredible! He hurried up the ladder after the other man and entered the ship.

Thymine examined the complicated network of controls just inside the entrance. He glanced quickly at the television screen on the secondary control panel. The flames were getting closer to the stockpile. "We must launch immediately."

He returned to the entrance, closed the heavy door, and locked it. Then he led the other man into an elevator that brought them to the main observation room. They watched the flames edging toward the superstructure of the vault.

"Sit down and fasten your belt," commanded Thymine. "Pull that lever." The other man obeyed. Thymine pulled two other levers at the same time. The ship grunted, and lurched. Within seconds the earth was more than a mile below.

"We need nine more seconds to escape the shock wave from the explosions." And Thymine began to count: "Nine, eight, seven, . . . two, one." Relieved, they slumped in their places and looked out the large window. Far below them was a tiny flash of light. Larger—reds, oranges, yellows, and whites. Smaller. A tiny speck. Earth no longer existed. Nothing. Nothing but space and light. They looked at each other, completely dazed, shocked.

After waiting for the automatic gravity compensator to start, Thymine detached his safety belt, lifted a hatch in the floor, and climbed down a short ladder into the control room. He turned around and saw at the controls the mannikin—a robotlike object with a man-woman face and no hair, an aperture like a mouth, other apertures like eyes, and beneath all that, a brain as complicated as man's.

"So here we are, my pretty robot, in your hands."

Silence. Then a hollow, metallic response, "Yes," like a deep voice at the other end of an interplanetary phone system.

Thymine stooped to examine the numbers and letters on the monitor underneath the chair of the mannikin. The data of their survival or destruction were being recorded before his very eyes in the constantly changing hieroglyphics of the computer. He wondered if there was a way to reprogram the robot. He sat on the floor, and recalled how long he had worked on this machine. It brought back to him memories of childish machines, high school rockets, college fuels, and finally the Technical Institute where he had gone to perfect the theories by which he would eventually bring man to the apex of progress and perfection—he, physicist, chemist, biologist, programmer, the prototype of what man could become through control of his DNA. His father and his father before him had planned things in such a way that he, Thymine, was the first of a race that would become increasingly more perfect and self-reliant, increasingly in control of its environment, its offspring, even death, then planets, and finally the universe. But none of that now. Instead, it had come to this—off into space fleeing rather than conquering. And with no other living creature than this Daedelus, a common product of centuries of his kind, a philosopher, a poet, an artist. And a robot, programmed to land on planet Terminus where nothing could survive beyond a few weeks. He smiled bitterly.

"Robot!" In a soothing, careful voice he addressed the object. "Can you go somewhere other than Terminus, the destination to which the Committee has programmed you?"

A long silence. Then the metallic voice. "Yes. Two places: Planet Laboratoria and Planet Bibliotheca."

Thymine decided immediately. "To Laboratoria then."

He was about to mount the ladder to the observation room when the metallic voice stopped him: "You and your companion must agree on a destination. Agreement will set off physio-intellectual-chemical reactions which shall readjust slightly the meteorological conditions of the ship. This change in turn shall automatically reprogram me to go where your unanimity indicates. If there is no agreement, there are two other possibilities. Either the ship will proceed to planet Terminus, or it shall reach a point hypothesized by the Planners and designated by the name Line of Emblance. It is not known if such a line exists, but if it exists and if the ship reaches it, something in the quality of the atmosphere shall affect me in a way about which the Planners knew nothing. I can say no more."

"But there is only one companion, and he is a poet, or philosopher, or some combination of these and other things that are useless."

No response. Only silence. Only the unseeing straight gaze of the mannikin, arms and legs outstretched to controls.

Thymine smothered a curse. Agree with a poet! He looked out the cabin windows and saw only light. Light could soon be dark, and dark could be light, and hot could be cold, and cold, hot. That's the way poets thought. His perceptions would be like the poet's: everything fused together. No distinctions. No clarity. No certainty. Only oneness and entropy and consequent indecision and immobility. To be like a poet? Death would be preferable.

He would explain. Certainly, even a poet would finally see the truth under these conditions. Thymine mounted the ladder and, as he reentered the observation room, noted that Daedelus was sleeping or thinking or dreaming. What did it matter! Let him continue. There was time enough for decisions, and revisions and indecisions which the poet would demand. And Thymine knew that somehow he had to get Daedelus to agree. He was thankful that his intelligence had been purified by the processes which his grandfather had perfected. There should be little challenge.

Suddenly loud static and a raspy voice on the intercom: *"one hundred twenty minutes."*

Daedelus awoke with a start. "What was that?"

Thymine responded lackadaisically, "Nothing but the robot giving the first of his time announcements," and looked hard at Daedelus. "You know," he said threateningly, "we have one hundred and twenty minutes to decide where we will go."

"What does that mean?" Daedelus was startled.

"It means this," Thymine went on to explain. "The robot has been programmed to go to two other destinations aside from the original destination of Planet Terminus, where we could survive only for a short time. On either one of the other two planets we could survive. But before the robot will be able to redirect the ship, we must agree on one of these two planets as our mutually acceptable destination." And he repeated the rest of what the robot had told him.

Daedelus looked at him with suspicion. "What are these two planets?"

"One planet is called Laboratoria. It is a place of great scientific achievements, material resources, and comfort, which my colleagues and I had been building. Computers are already installed and now are receiving power from atomic generators that were activated during our expedition there last week. A dome thousands of square miles in area has been successfully suspended over the most fertile part of the planet, and rain falls periodically from the condensation of water vapors generated from the reactor's heat exchangers. Specially adapted seeds are already sown, luckily, and animals bred. The place is a veritable paradise, a dream beyond the wildest dreams of scientists of former days. It is not yet completely done, but with assistance of third generation robots already there, whose strength and intelligence exceed that of any man of past ages, I could finish it in a short time, and could spend the rest of my days doing pure research, utilizing the most advanced facilities man has ever known. I, of course, wish to go there."

"And the other planet?" Daedelus was thoughtful and interested now.

"The other planet has been named Bibliotheca. It is full of earth's books and manuscripts—nonscientific and nontechnical garbage of course—that were deposited there by a stray ship originally intended for the vaults of Laboratoria as decreed by the Secret Committee of the Inner Core. Although instruments indicated that the crew is dead, probably from the impact, they also indicated that the environment of the planet is naturally conducive to human survival; there is warmth, water, vegetation, and according to television monitors, some kind of animals, the species of which I know nothing. Aside from these factors, there is little of interest. The records that the ship contained are of no value, as they are but evidence of human folly from the beginning—theology, philosophy, literature, art, and so on. It is the kind of place which the inferior early part of our species might desire—poets, for example. It is—how shall I put it with delicacy—the sort of place you might choose."

"Quite," replied Daedelus. "Quite. What do I care for progress when it is progress that has destroyed earth and brought us here."

Thymine reddened in anger, but considered a sharp response beneath him. "My friend, our only chance is to agree upon a destination. Let us, then, put aside our differences and discuss, until we arrive at a common decision. Certainly two rational and intelligent men like ourselves can do that in whatever time we have left." Daedelus slipped out of his suit to be comfortable during the exchange of ideas that was to follow.

Thymine pondered the slowness and dullness of it all. He, the prototype of the new man, would have to revert to the witlessness of his forefathers to accomplish his survival. The humiliation!

Daedelus looked at his companion. If scientists had listened before, the two of them would not be here. Progress! The opium of the scientist!

Wordlessly they both went to the automat, drew cups of coffee, and returned to their seats. They sipped, preparing themselves for the dullness that each expected the other to visit upon him. The ship was silent except for the whoosh of the gravity compensator and the hum of the computers.

"You, my friend, may begin." It was Thymine speaking.

Daedelus cleared his throat. "Will you agree, Thymine, that our destination should be that planet where our being has the means which are most likely to perfect itself, to make it happy, to ennoble it?"

"Of course." Thymine yawned knowingly, as if he had heard all this before.

"And will you agree that such things are determined by what man's nature really is, for every being seeks that which is natural to it, and that which is natural to it is that which perfects it or makes it basically happy?"

"Yes."

"Well, then," and Daedelus took a sip of coffee, "since man is obviously a being comprised essentially of matter and spirit, we should choose that planet which most likely will provide the means of perfection, or happiness, or ennobling of these two parts of ourselves, namely, the planet that provides both happiness to body—which is good health and comfort—and happiness of spirit. But most of all happiness of spirit, for it is spirit or soul that lives on, that makes man what he is, that distinguished him from the rest of the animals that used to exist on that planet which your contrivances and discoveries and progress have just managed to annihilate. In view of this, it would seem sensible for us to choose Bibliotheca, because there are to be found the means of our perfection, our happiness, our nobility, the great works of art and literature and philosophy, the records of history, and even the works of theology, which, I'm sure, you have always considered to be next of kin to superstition."

"But, as is usual with people like you," Thymine interrupted, "you emphasize some fuzzy concept of man's spirituality and downgrade man's rational nature and his body."

Daedelus ignored him. "To be more specific, man is composed of body and soul. The discussion of body, though, is of no real import here or anywhere, for even the lower animals had body. So let us concentrate on soul, or spirit—it is hard to decide what word to use for so subtle a concept. Soul is composed really of two faculties, intellect and will. These are the faculties that make man distinct, which make man to be man, which back on earth used to separate him from the other animals. Thus, what perfects these faculties, or what gives them happiness, or what ennobles them will make man himself more perfect, more human. Now it is clear that what satisfies intellect is truth, and what satisfies will is the possession of what it desires most, which is good. So, Bibliotheca should be our destination, because it contains the means by which we can achieve truth—namely, books of philosophy, of history, theology, and so on—and because it contains the objects of art and literature in which we can achieve what we at bottom desire (at bottom, because we desire many harmful things under the guise of good); that is, the truly good, the truly pleasurable, the truly ennobling—words difficult to use because the reality they signify is so profound and elusive."

Daedelus settled back in his chair contentedly and authoritatively.

"Incredible," Thymine muttered.

"Pardon me?" Daedelus rose to the edge of his chair.

"I said *incredible*." And Thymine's face reddened from the effort of the shout. "How in the world do you know what man really is? What is spirit? What is soul? Have you ever seen them? Can you measure them? Predict their behavior? If not, then you simply cannot have ever known them and do not know what you are talking about."

Daedelus took out his pipe and lit it with deliberate calm. He was hiding anger. "My good man, I, the lover of the human, if you must know, see clearly what man at bottom is. This knowledge of mine is part of my carefully cultivated mystical oneness with the universe. As I have several times indicated, had the race listened to the lover of the human, to the true philosopher, the true poet, the true artist, you and I would not be here now, without hope of progeny, without hope of anything except perhaps mere physical survival. And the race would not now be extinct except for you and me. It is your kind, after all, that helped man progress from atom bomb, to hydrogen bomb, to cobalt bomb, and finally to n-tron bomb. So it is certainly not you who have known man. Otherwise, you would have ignored that which obviously was not good for man in any way. If for no other reason, I am competent by default."

Arising in obvious agitation, Thymine said, "I cannot help it if the fools who ran the governments of earth failed to understand and thus misused what science had provided for man."

Daedelus broke in. "Failed to understand! Misused! Ha! A likely excuse and one I've heard many times before. You're all alike, and how well you always clung to each other. Did you and your kind stand up to the governments? No! You were too interested in amassing financial assistance from the state, ostensibly for public health. But did you use it for this? No. You developed strains of disease-causing microbes which were resistant to all known antibiotics. You turned public health upside down, and if your bomb had not destroyed the race, your public health would have done the job! In other words, you were interested in your own prestige and the prestige of your institutions of learning, too concerned with progress to worry about truth and values. Your truth was measurable, was seen in microscopes, was computed, was limited to the predictable, the manageable. Truth, real truth, the truth of what it is to be really human, the truth of what the human really is, is accessible only to the mind of the philosopher, the poet, the artist, who was, in the civilization you were developing relegated to the junk heaps of the world, or else transported to the vaults of your cold, artificial planets. Days were when the theologian was ruler of those who claimed to know. Even that was preferable to the last age of the earth. In fact, any other age of man was preferable, as a sampling of his achievements reveals—the Iliad, the Parthenon, the Venus de Milo, the Coliseum, the Divine Comedy, the Cathedrals of the Middle Ages, the plays of Shakespeare, the magnificent and finely woven fabric of the Kantian system, the poetry of Goethe, the canvasses of Van Gogh, and so on. These achievements are all symbols of the great ages."

Thymine had had enough. "Yes. And look at the slaughter on the plains of Troy which your poets needed for inspiration. Look at the glorious past— the Gladiators, to say nothing of the early Christians, in the Coliseum. Look at the pillage and suffering of the dark ages, the brutality of so-called knights, the terrors wrought by the ministers of religions. Look at the role given to women. Look at the misery in which nine-tenths of the world lived for thousands of years until the development of science and technology. Look at the ruins of mind and body that your superstitions have strewn over the centuries. And look too at your own delving into the coffers of the state or at your own flattering of patrons so that you could paint your pictures and write your novels, while the rest of humanity was starving, disease-ridden, illiterate, and killing one another in the streets and on the fields of battle. Look at what you considered to be the perfection of man, a perfection that taught him to seek only the things of the spirit, because you were too proud to admit that you knew nothing of the things of matter. See what suffering and destruction you and your kind visited upon the race, not only by commission but also by omission. My friend, before you throw stones, examine the glassiness of your own house."

"*Ninety minutes.*" During the quiet that followed the time announcement, both men grew thoughtful. Daedelus looked up and responded in a surprisingly soft tone. "I have had my say. Speak your mind!"

Thymine hesitated a moment, and replied, "Thank you, my friend. If I seem thoughtful it is because what you have said raises several important issues." He walked over to the automat and refilled his cup. He returned, looked at the ceiling for a moment, and began to talk.

"It seems to me, Daedelus, that you have been arguing in a strange way. First, because you are philosopher, poet, artist, in general, lover of the human, you automatically assume that you know what man is. Second, because you know what man is, you conclude that you know what is good for man. Third, because you know what is good for man, you conclude that you, the lover of the human, should decide on our destination. Naturally, you prefer the planet Bibliotheca, as I knew you would."

Daedelus looked up. "Although your analysis of my argument is accurate, I do not agree that the way I argue is strange."

Thymine went on. "That is your privilege. But let me continue to reveal my own positions on these matters, and in revealing them try at the same time to undermine some of the things you have said."

Daedelus removed his pipe from his mouth, "I've been waiting. Be my guest." He was perplexed at Thymine's earnest softness.

"First, Daedelus, I would like to make clear my position on which one of us has the right to speak about what man is. For reasons which you have not at all made clear—I believe you attributed your authority to a mystical oneness with the universe which you and your kind supposedly enjoy—you say that the lover of the human is *the* authority on what man is. This seems to imply that I, the scientist, do not love the human. It also seems to imply that even if I do love the human, it would have to be a human different from that which you recognize and claim to love."

Daedelus nodded in agreement.

Thymine continued. "I want to assert that I too feel a oneness with nature, which, in fact, I would go so far as to say is my religion. But my credo is one of change, not dogma. By observing, experimenting, hypothesizing, theorizing, investigating, reinvestigating, yes, by utilizing the methods of science, I and my type gain a better understanding of nature and thus man. From this continually increasing body of knowledge springs my feeling of unity with nature and love of man. You and your kind have been repelled by science's discussion of man because you cannot understand it, or better yet, *will* not understand it. The phrase they used to employ was, I believe, communication gap. There was, you must admit, such a gap between scientist and so-called humanist. You and I perfectly illustrate this gap."

Daedelus was a bit annoyed. "I am tempted to respond that if such a communication gap exists, then it is due to the arrogance of scientists, who have maintained that any problem can be treated scientifically as long as it is stated in an intelligible way. But proceed. I have had enough of violent controversy today. After all, it is our mutual survival that interests us."

"Well spoken, my friend," Thymine rejoined. "Would it help any if I say that my kind have often tried to insist that theories are forever tentative?"

"Yes, I believe it would." Daedelus was impressed by Thymine's modesty. "In response to your conciliatory attitude, Thymine, I will admit that if there have been scientists who maintained that science alone can and does achieve certainty, there have been many so-called lovers of the human who have assigned the same kind of foolish claims to their own disciplines. For thousands of years, for example, there were those who maintained that metaphysics was the queen of knowledge. And yet these same men, for metaphysical reasons, argued that the earth was the center of the solar system."

Thymine answered, "Yes, and all because they felt man to be the center of the universe."

Daedelus rose for more coffee. "That is a point which I am afraid I shall find hard, perhaps impossible, to relinquish."

"Well, let me proceed." Thymine was eager now, perhaps even confident.

"Certainly," Daedelus returned to his seat.

Thymine went on. "First, I must say that I see man as only part of the universe, closely joined to other living organisms and even to the inorganic world both by necessity and by a generalized concept of evolution consisting of physical, chemical, biochemical and biological phases. Thus man is only one among many other material entities. This is the unity that I referred to a moment ago. Let me elaborate on unity in biology. A large number of experiments has established, beyond doubt, that all earthly organisms, whether they are microbes, plants, or animals, including man, carry out the same fundamental processes. All transform energy from their environment into a product we have designated as ATP, and then use the energy stored in ATP to perform biological work. All utilize DNA for genetic material, and are replicated by a similar mechanism. All express their inheritance through the production of enzymes, which are synthesized on ribosomes, utilizing messenger RNAs as templates. This common theme of life is spectacular evidence for the theory of evolution. All this is to say, of course, that to deny that man is just one animal among many others is to deny what is obvious. For those who believe that man is created specially, I can only quote (who was it—Shakespeare?), 'Vanity of vanity, and all is vanity.' "

Daedelus' response was quick. "The very way in which you speak tends to corroborate my feelings that you are responsible for the communication gap. And besides, if we go to Bibliotheca I can show you the place in the pages in the Book of Ecclesiastes (I: 2) where the sacred writer utters his famous epigram about vanity." His tone was a bit surly.

"Manner of speaking? You criticize the shortcomings of my speech?"

Thymine was again angry. "Look at some of your fool poets, and then criticize me for not being clear in my speech."

Daedelus, comforted by his success in breaking down his companion's composure again, waved his cup cavalierly. "Continue with your exposition, or we shall never decide where we are going."

Thymine was tempted to do something drastic. Vaporizing the other man entered his mind, but he was afraid it might destroy the possibility that somehow, though the outlook was becoming more bleak, the robot might be reprogrammed away from Terminus to one of the planets where survival was possible. So he controlled his urge. "Aside from difficulties of language—and you must admit that poets have not always written clearly, nor have artists always painted things intelligibly—you must realize that to understand man one must understand the chemical nature of a cell. Everyone, with a little work, can understand this. So let us not quarrel about the quality of our speech. We have more critical matters to deal with, like ordinary communication, so that we may survive."

Daedelus, feeling a bit foolish, shifted uneasily in his chair. "I am sorry. You're right. Though I am a layman when it comes to the matters you have just discussed, I admit to a rudimentary knowledge—obtained while submitting to the university's requirement that I take science courses—which enables me to follow your exposition. Please go on."

"So you are not completely unfamiliar with matters scientific. I too confess to a familiarity with matters humanistic; music, for example. Thank you. Perhaps the gap between us is lessening. I had discounted both your intelligence and our ability to ever agree. I may have been vain, but worse, over-pessimistic." And Thymine took another sip of his coffee.

Daedelus took advantage of the quiet and commented, "Let us at least agree in the hope that your pessimism—and mine too—was unfounded." He gazed out the observation window and saw strange crystal fragments. But before he could inquire about them, his companion was speaking again.

"Now to my second point. Because man is only one material entity among many, it seems to me that he who has the knowledge for and the method of deciphering matter is the one to decide what is best for man, thus which planet is best for us. I, who have such knowledge, feel that we should go to Laboratoria, for there we may find not only all the means of achieving our well being, which is our material well being, but there we shall also have at our disposal the means of learning more about ourselves, about the rest of reality, about how to achieve further progress. For, on Laboratoria, in addition to the advantages which I briefly pointed out earlier, there are thousands of programs ready to be fed into computers, programs that will extend knowledge about ourselves beyond anything that has been known. We may even discover the means of conquering that which has terrorized man since time immemorial: death. Think of it, to discover the means of achieving immortality; that is, immortality of body."

Daedelus was deeply perplexed by Thymine's last comment about death. "It is hard for me to conceive of deathlessness in your terms. I have never really feared death because I always knew that my spirit, my soul, would live on—perhaps wandering pleasant places like the Elysian Fields of the ancient Greeks, perhaps moving into some other form of being, as the Hindus suggested, perhaps even enjoying a place some called Heaven. Why should I want to go where I would become obsessed with your kind of immortality, that of body, when I already possess immortality in my very nature? 'Not by bread alone does man live' Deuteronomy (I: 3). In short, I have never lived with an eye toward survival in your material terms."

Thymine was quick to respond. "I do not think your position regarding death is as uninterested in body as you think. Let me say this about your attitude toward death. It seems to me that man's desire to propagate, to leave behind part of his body—genes—in his offspring, certainly reveals a basic human attempt to avoid total material extinction. And it seems to me that the desire for immortality of some kind is revealed also in the labors of poets and artists, who would see themselves remaining materially, in some strange way, on the printed page or on the painted canvas. It seems to me that philosophers and theologians revealed an attempt to avoid total extinction in the systems which they elaborated, systems which they hoped would have immortalized them by putting them on the lips of all men for all times. So it seems to me that the lover of the human, like the scientist, has also revealed a desire to live on tangibly in some way. Finally, it is natural for one, faced with the grim prospect of death, to search for some way of alleviating the fears arising from this prospect. In ages past, man's body died: it was as simple as that. So it was natural for man to discount the importance of the body, about whose death he could do nothing, and to alleviate his fear of non-being by finding something in him that could not die, namely, soul or spirit, as you have said. So I ask you to consider the possibility that the kind of immortality you subscribe to is a fantasy which man's need has converted to doctrine in order to protect himself from both thoughts of death while he lives and the reality of death when it is imminent. But if, as I think possible, we may discover the means of conquering death, then you no longer need the fantasy of your kind of immortality."

Daedelus was very thoughtful.

"And furthermore," Thymine continued, "your ideas on death resulted in an attitude toward the body that seems to have been largely responsible for man's doing so little to control his environment, to better his lot, to improve the quality and the length of his earthly life. Yet certainly you have profited from the achievements of science. You are a man of fifty-five or sixty years of age. According to statistics, if you had been born in 1850, your "average expectation of life" would be 34 years. A baby born in 1950 could expect to reach an age of about 65. Today, you can expect to live for 85 to 90 years. In other words, if it were not for science, you would be, as the insurance companies used to say, 'actuarially dead.' If science has done nothing else, it has given you, the lover of the human, a longer time in which to discover more about the nature and end of your existence, and has given you greater comfort while pursuing these questions."

Daedelus answered, with melancholy and perhaps regret. "Much of what you say may be right, but your hope to conquer death, and to create life, although you have not spoken of this, by the scientific and technological means present on Laboratoria, is still a mere hope. In fact, I see it as a pipe-dream, the same kind of sublimation of which you accused me. Nevertheless, I do admit that the modern lovers of the human really did not use their added days to do much except persist in the same age-old petty quarrels about such questions as whether man is intrinsically good or intrinsically evil; about whether man should or should not eat, drink and be merry for tomorrow he would die; about whether man should or should not wallow in the infertile solipsism of his individuality, which, I am sad to say, man did in the last age to such an extent that he would deny not only others but even himself the power of ever communicating because he could never in his view escape the all-consuming self. Such solipsism is certainly the most genuine substitute for suicide, of which Camus was always fond of talking. Such attempts to escape into the self probably explain why much literature of recent ages babbled absurdly, why much painting was full of confusion, why much music was unpleasant to the ears. The lover of the human, for better or for worse, certainly does not have a history of progress."

Thymine was consoling. "If the lover of the human scorned progress and thus scorned man, please be aware that the scientist was often guilty of a selfishness just as extreme. Your rebuke of the scientist for his cooperation with governments in the development of the bomb is legitimate. His refusal to cooperate with the state might have eliminated the catastrophe which we witnessed minutes ago. Perhaps there is something in man, whether humanist or scientist, that makes him destructive of himself."

Both men were very silent now. Each one was ruminating about the childishness of his alleged superiority over the other. It seemed to each of them that to be man was bad enough, but to be a man convinced that he has the truth, and he alone, was even worse. Such an attitude could result only in extinction, when it was survival they were pursuing.

"*Sixty minutes.*" The hollow metallic voice again echoed the time.

"It is possible,"—Daedelus was speaking—"that your kind and mine have committed themselves to the same preposterous position, namely that truth is distinct and one, that truth is exhaustible. Such a position would lead us to conclude that there is one method of seeking this one truth, and would send us off in search of this method. Having deceived ourselves into believing we had found the one method, we would then go on, like a horse with blinders, pursuing our own visionary course. You would have proceeded further and further from the concerns of art and literature, from the valid insights of philosophy, because none of these things provided data which you could measure and isolate and put into a formula. And I would have gone on denying the merits of the discipline and qualified kinds of truth that science offered and used to improve the lot of the race, to increase knowledge of ourselves, and that of the world around us. We each would have gone on propagating our species, now not human, but species 'scientist' and species 'lover of the human.' And had there not been war between Eutronia and Aggressia, there would have been war between your kind and my kind, though my kind, with horse and spear like the Man from La Mancha, would have met with inevitable destruction. And your kind would have been the worse for our extinction, as mine would have been the worse for your extinction." He paused a moment, and asked ruefully, "Thymine, why did you ask me along, when you knew who I was?"

Thymine thought a moment. "Because, I suppose, I have my doubts about what man really is, and because, especially, I was afraid that the absence of human companionship might actually be far more fearful than the prospect of material immortality was desirable."

Both men stared quietly out the window.

"Daedelus, I suppose our differences come down to the question whether man is only body, or whether he is only soul. And I'm afraid that I stand rooted firmly in the conviction that man is only one material entity among others."

Daedelus responded, "And I stand rooted firmly in the conviction that man is matter and spirit, and that we—each a man—are better off where we have available the things that feed the spirit."

Thymine was dejected. "But don't you see that you are not saying that man is both matter and spirit. Rather you have really been saying that man is spirit."

The possibility of compromise that Thymine suggested had already entered Daedelus' mind, but was removed by a sudden topsy-turvy motion of the ship. And instead of following the thread of compromise, Daedelus was frightened and totally distracted. Grabbing a rail for support, he shouted in fear, "What was that?"

"Just a small meteor colliding with the ship. I suspect that we shall either go to Terminus, or shall reach the Line of Emblance, for it seems unlikely that we shall come to a mutual decision that would reprogram the robot to either Laboratoria or Bibliotheca."

Thymine went to the automat and returned with a number of very small capsules. He handed some of them to Daedelus.

Daedelus, still a bit frightened, took them without thinking, and then asked, "What are these?"

"Lunch," the other man responded, and popped three capsules into his mouth. "Take them. They contain all the nutrients and vitamins that you need."

Daedelus obeyed, but then commented with a dejected, if conciliatory tone, "This is the kind of thing I have detested most about what science has designated as progress. Scientific progress has removed all the pleasures from life, even the simple pleasure of eating."

"I suppose you are right, my friend. But progress is pleasure. What you and your colleagues have always failed to understand is the difference between science and technology. Science is not building bigger machines or better gadgets although they play a role. Those material items to which you refer are but commercial by-products. The primary concern of science is ideas. In this regard we are akin. But whereas you glory in the past, science is more interested in the present and future. Where your ultimate authority is in the classics, mine is in the latest experiments. These new data inspire fundamental changes in the conceptual framework of science. The ideas of science have permeated the whole cultural pattern of Western Civilization. Who is to deny the philosophical and social impact of the ideas of Pasteur, Darwin, Einstein, Freud, and Watson, and Crick? Given time, the ideas of Ramasarma, Zotov, Ling and Mo-Jo, Eisenberg, and Di Marco also would have changed man's philosophical outlook on the world."

Daedelus shook his head slowly from side to side. "You realize, of course, that it is hard for me to see this as progress. But I would willingly listen to a presentation of other conquests which you think science has made. I am, by necessity, at the point where I will try hard for compromise, and I think you are too. It is the instinct for survival operating now, which your kind and mine have always understood in pretty much the same terms. We can at least say this much about ourselves, that the threat to survival does not seem capable of destroying the integrity that each of us has."

"Your observation about our integrity is somewhat comforting, but before I go on, let me first take care of some necessities which science has not yet done away with." Thymine disappeared into a small door, and emerged a few minutes later. "I have already pointed out that science aligned with technology increased man's life span, and may soon provide you and me with immortality. It also improved the quality of man's life by making him less the victim of discomforting and debilitating illnesses and by improving the comfort of his surroundings on earth and his control over his environment."

Daedelus thought for a moment. "But on the other hand, your advances, or if you insist, the commercial by-products of your ideas, were responsible for many unfortunate occurrences on earth. Passing over the folly of the bomb, which both of us admit, I can point to any number of ways in which science's dabbling in the mysteries of nature caused the balance of nature to be destroyed. For example, pollutants from the chemicals of industry ruined most of earth's rivers and streams, and made it necessary for the occupants of most cities to wear special breathing apparatus. Tampering with pollens of every kind was responsible for allergies that developed at a rate faster than the scientist could keep up with. The elimination of infant mortality and increased longevity brought about a population explosion and resulting food and space problems that most certainly would have been impossible for even the advanced technology of the last age to solve in time to avoid war. But worse than all this, science had developed a type of human being in large and constantly increasing numbers that was in almost every way little different from the advanced robot driving this ship."

The metallic voice, as if in rebuke of Daedelus' uncomplimentary remark, echoed, "*Thirty minutes.*"

"Damn that thing," they both shouted in despair.

Thymine wanted to speak, not angry anymore, just anxious, but settled back in his chair, hopelessly, to listen.

Daedelus continued. "In effect, science was perfecting a kind of man that centered his every effort on the material, had brought him to a point where it was impossible to enjoy the rapture of a great symphony, the soul-searching experience of a profound drama, the insight of a great painting. Science, in other words, had almost succeeded in making man believe that he is only matter, only body, and that only things material and things bodily are of consequence. If I am not mistaken, the scientist had even come to see love in terms of the expression of specific genes, and science was attempting to improve these specific genes to a point where their base composition would be ultimately responsible for a higher order of love than man had ever known. In all honesty, respected companion, I must, with great dejection, admit that such ideals and such achievements are indeed beyond my ability to condone and, even less, to desire. If Laboratoria is the kind of place set up for these kinds of ideals, then I, whether I am what I have made of myself, or whether I am the product of generations of my kind, certainly have no wish to go there. Death would, I think, be preferable. At least I die with my hopes for immortality, whatever it may be. The unknown seems more desirable than what I have seen man becoming, and what I fear I would become on Laboratoria."

Thymine put his elbows on his knees and stared at the floor for some minutes. "You have spoken eloquently, and with great feeling, poet, and I have come to respect you."

"Then the feeling of respect is mutual," Daedelus responded.

Thymine continued. "Nevertheless, I think you are here revealing the brooding self-pity and consequent despair in which the lover of the human always seems to have immersed himself. You seem to be losing not only your capacity for clear thought and decision, but even your desire for thought and decision. You seem, like many writers of the past—I think of Shelly, Arnold,

Camus—to be surrendering yourself to immobility, to confusion, to a sort of solipsistic and sterile contemplation to which lovers of the human of all ages seemed to have fled in order to escape imminent peril of mind or body, or both. Is there not some point at which you might meet me? The only thing I must hold is my unshakable conviction that man is finally the composite of distinguishable particles of matter. If you can acquiesce in this, then we can certainly agree on other issues. For I am certain that I can help you realize that time and progress would have removed the shortcomings of which you accuse science—pollution of natural resources, population problems, and so on. Furthermore, I am certain that the duality of mind and body to which you constantly refer is only our temporary ignorance in the young sciences of psychology and sociology. Eventually, there will be a molecular explanation of these problems.

Daedelus, hopeless and full of despair now, slumped in his chair. "But science cannot solve such things as the problem of existence and state clearly its purpose. Nothing can do this. Such problems are in the realm of mystery, the reality of which I affirm and you, in your very being and words, deny."

Thymine responded, "You can't possibly believe in mysteries. That's like believing in miracles. Both mystery and miracle are words which man's pride has given to that which he does not yet know or understand. Affirming the reality of mystery is like affirming the existence of God. And can you prove his existence?"

Daedelus looked beyond Thymine and noted to himself that the light outside was increasing in intensity. "I suppose you wish me to provide a formula which shows God as measurable, as this or that, and so on. No, I cannot do that. But then I must ask you these questions: can you prove to me that mystery does *not* exist? that miracles are *not* possible? that God does *not* exist? If you respond by saying only that exists which can be measured, then we are back where we started some minutes ago, after the other catastrophe, for I believe in a kind of knowledge that cannot be expressed in terms of measure of any kind. I believe, in other words, in a knowledge that transcends scientific knowledge and that is finally not only more complete because it is multi-faceted, and thus in harmony with the many-sided quality of reality, but also is more accurate because it is immediate, intuitive; that is, my mind and its object are one. It is the kind of knowledge that is like my oneness with the universe."

"Thirty minutes."

Thymine had arisen from his chair and, ignoring the metallic voice, was about to respond to what he considered to be useless mysticism. But suddenly there was a brilliant flash, and a force like an explosion. The ship began twisting and turning, end over end.

"My god," cried Thymine, "it must be Line of Emblance! The ship's time mechanism was probably destroyed when it collided with that meteor."

Both men, grabbing their seat belts, were being swung about like a nail at the end of a whip. Each time the ship twisted and turned they were bashed against walls or some other objects in the room. There was another jolt from the unknown force, and both men were thrown against a metal cabinet that stood in the corner of the observation room. Then, as suddenly as the twisting and turning had begun, it ceased. The ship righted itself, and the bodies of the two men, bloody and unconscious, slid into the middle of the floor as it started on a new course. The room was a shambles of papers and cards shaken loose by the violent movement and twisting. There was silence. Nothing moved inside. But the ship continued on through rainbows of brilliant colored light.

After what must have been an hour, the hatch leading to the lower cabin and control room popped open. Then two hands with long white fingers and a head emerged—a woman's head with long blonde hair and lovely face— and then her body, clothed in what had been the covering of the robot. She looked around the observation room, saw the two men, and went over to them. She untangled them from each other, went to the automat to draw water from one of the tubes, and treated their wounds. Then she went to the metal cabinet in the corner and took from it a hypodermic needle and a small vial. She inserted the needle into the vial, and watched the vacuum tube fill with a blue fluid as she pulled the syringe back. She injected the fluid into Thymine's arm, then Daedelus', and sat back to wait for the effect.

Daedelus was the first to revive. He raised himself up on one elbow. Then Thymine awoke, and sat up, like one just awakening from a sleep. They both saw the beautiful woman at the same time. Daedelus tried to speak, but no sound came to his lips; Thymine also could not talk.

The lovely woman walked toward the automat, turned around, and then spoke. "Gentlemen, you cannot talk because we have passed the Line of Emblance. It will be some days before your powers of speech will return, but do not be concerned. You are both fit, and will speak normally. Let me try to answer the questions that are undoubtedly disturbing you."

She then went to Thymine, and helped him into a bed which she pulled down from the wall. She did the same for Daedelus. Then she pulled a chair out of the rubble, and sat down.

"First, know that I am Venadeninus. I was, until a few moments ago, the robot in control of this ship. As the Committee had hypothesized, there *is* a Line of Emblance, which turns out to be a thin belt of electromagnetic radiation of greater intensity than even you, Thymine, could have envisioned. And whether the ship is still intact due to the will of a deity or because of man's skill we shall, I'm afraid, never determine. In either event, these physical forces penetrated the control cabin, but did not penetrate this observation room because of the plastic substance which insulates the walls. These rays were responsible for turning me into a being that seems to be human, like yourselves."

At this, Thymine motioned for a pen and pad. He was too weak to get up for it himself. Venadeninus discovered a pen in the rubble, picked up a piece of paper that was at her feet, and handed them to Thymine. He wrote out a message and gave it to her. She looked at it, and then tore it up.

"Although I cannot read your message, I know what it says. You ask where we are going, do you not?"

Thymine nodded.

A playful smile crossed her face. She rose, went to the hatch, disappeared into the lower cabin, and returned a moment later. The mischievous smile was still there. "I suppose," looking at Daedelus, "that you wonder about the same thing." She sat down again.

"There is no need to worry about Terminus anymore. Nor shall you have to worry about Laboratoria, Daedelus. Nor you, Thymine, about Bibliotheca. The forces of the Line of Emblance have put the ship on a course that will soon bring us to a safe and suitable planet, the name of which is Vita. It is a place where you, scientist, can start again in facilities that you will find useful and advanced enough, and a place where you, poet, will find books in plenty, and environment conducive to writing and thinking, and beauty enough to keep you painting and singing the rest of your days. It is, in short, a planet where we shall be able to start everything all over again. You should be grateful for the opportunity to finally reconcile your age-old differences and to learn to understand and appreciate each other."

The two men looked at each other for a moment. Then Thymine sat up a little, reached into his pocket, found the last capsule, popped it into his mouth, and chewed it without even noticing its terrible taste. Daedelus took out his pipe, and began puffing, without realizing it was unlit. Before they fell into a peaceful sleep, both men, though justly concerned with what the new planet would be like and what they would do there, were more interested in this strange being. Coincidentally, a line from the pen of a certain Robert Frost occurred to them: "Earth's the right place for love." Even Daedelus fell asleep entertaining the hope that perhaps Frost might have been wrong.

GLOSSARY

Exactness cannot be established in the arguments unless it is first introduced into the definitions.

HENRI POINCARÉ

Abiogenesis. See Spontaneous generation.

Active site. That part of a given enzyme molecule into which the substrate fits.

Adenine. A purine molecule found in both RNA and DNA (see Figure 1.11).

Adenosine triphosphate (ATP). A molecule consisting of adenine, ribose, and three phosphate groups. (For chemical formula, see Figure 3.4.) It is the key energy-rich compound in the cell, that is, when it is broken down a large amount of energy is released. Loss of one phosphate converts ATP to a diphosphate, ADP; loss of two yields the monophosphate, AMP.

Aerobic cells. Those that utilize oxygen.

Agar. A relatively inert polysaccharide derived from seaweed; it is used widely in microbiological laboratories as a solidifying agent in growth media.

Alkaptonuria. A hereditary disease in which the patient lacks the enzyme for degrading homogentisic acid and thus excretes it in his urine.

Alleles. See chromosomes.

Allosteric protein. Proteins whose three-dimensional structure and biological properties are altered by the binding of specific small molecules at sites other than the active site.

Amino acid. A group of organic compounds which comprise the building blocks from which proteins are constructed. (For general chemical formula, see Figure 1.10.) There are 21 different amino acids that are commonly found in proteins.

Anaerobic cells. Those that can live without oxygen.

Angstrom (Å). A unit of length equal to 10^{-8} cm.

Atoms. The smallest units in which elements can exist and still have their characteristic properties.

ATP. See Adenosine triphosphate.

Autoradiography. A method utilized for locating specific chemicals within the cell. A photographic emulsion is placed in contact with cells or sections of cells which have been made radioactive with specific components. The black dots (silver grains) which develop when the radioactivity is exposed to the emulsion point out where that chemical is concentrated.

Autosome. Chromosome other than a sex chromosome.

Auxotroph. A mutant microorganism which can grow only if supplements are added to the minimal medium.

Bacteriophages (phages). Viruses that infect and multiply in bacteria.

Biochemistry. The study of chemical substances and chemical processes of living things.

Biosynthesis. The process by which cells build large and complex molecules from small and relatively simple components.

^{14}C. A radioactive isotope of carbon with an atomic weight of 14; it has a half-life of 5700 years. It is a common tracer in biological and biochemical experiments.

Calorie. A measure of energy; the Calorie is defined as the amount of energy required to raise 1 cc of water from 14.5 to 15.5°C.

Cancer. A group of diseases characterized by uncontrolled cellular duplication.

Carbohydrate. Organic molecules having the general formula $C_n(H_2O)_n$, including sugars and polysaccharides.

Catalyst. An agent that increases the rate of a reaction, itself emerging unchanged at the end of the process. Since it is continually regenerated, a small amount of catalyst can produce a large increase in rate. In living organisms the important catalysts are large protein molecules called enzymes.

Cell-free extract. A solution which contains most of the ingredients of cells, prepared by breaking open cells and removing any remaining intact cells. One technique for preparing cell-free extracts is to place cells in a mortar and gently grind them with finely powdered glass; its abrasive action ruptures the cells and releases the cell sap. But if performed carefully in the cold, it does not destroy the DNA, RNA, enzymes, and ribosomes in those cells.

Cell wall. A relatively rigid structure found outside the cytoplasmic membrane and giving mechanical support to the cell; it is characteristic of procaryotic cells and plants.

Chemical evolution. The theory, independently proposed by Oparin and Haldane, which states that life arose gradually from nonliving material. The theory emphasizes the need for a long series of chemical changes as a prerequisite to the formation of life, with each change closely related to the one which went before.

Chemosynthetic bacteria. The group of bacteria that obtain energy by the oxidation of inorganic molecules and then utilize the energy to fix carbon dioxide into cellular material.

Chlorophyll. A green pigment located in chloroplasts and having the property of absorbing light energy and passing it onto other molecules.

Chloroplast. Chlorophyll containing cytoplasmic structures found in green plant cells; they are the sites of photosynthesis.

Chromosomes. Threadlike structures containing deoxyribonucleic acid and protein which are present in the nucleus of all animal and plant cells. They constitute the heredity organelles of the cell. In higher organisms the full complement of genes in a cell may be divided among several different chromosomes. In addition, each chromosome may occur twice (diploid), so that there are two pieces of genetic material controlling a given characteristic in any one cell. These two so-called alleles are the two "factors" postulated in Mendel's laws. In bacteria all of the genes are located on a single chromosome.

Citric acid cycle. See Krebs cycle.

Clone. A population of cells all derived from a common ancestor by asexual reproduction.

Coacervate. A "dynamic molecular aggregate" that Oparin postulated played an important role in the origin of life. A coacervate is formed when groups of large molecules such as proteins associate to form microscopic droplets in liquid media.

Codon. A sequence of three adjacent nucleotides that code for an amino acid or chain termination of proteins (see Genetic code).

Colony. A population of cells growing as a compact mass on a solid surface, usually descended from a single ancestor.

Compounds. Combinations of two or more different atoms held together by chemical bonds (derived from the Latin words meaning "to put together").

Conjugation. A physical association and exchange of genetic material.

Constitutive enzymes. Those produced by the cell at all times, irrespective of growth medium.

Crossing over. In genetics, the process of exchange of genetic material between homologous chromosomes.

Cytology. The branch of biology exploring the structures and functions of cells.

Cytoplasm. That part of the cell inside the cytoplasmic membrane but external to the cell nucleus.

Cytosine. A pyrimidine molecule found in both RNA and DNA (see Figure 1.11).

Dark reactions. Those parts of the photosynthetic process which can proceed in the absence of light: the formation of glucose from carbon dioxide and water.

Death phase. That part of the growth cycle in which the number of viable cells decreases sharply.

Degenerate codons. Two or more codons which specify the same amino acid.

Deletion. In genetics, loss of a section of genetic material from the chromosome.

Deoxyribonuclease (DNase). Enzyme catalyzing degradation of DNA.

Deoxyribonucleic acid (DNA). The chemical substance which is the genetic material of all cells. It is a polymer of deoxyribonucleotides. (For chemical formula, see Figure 5.4.)

Deoxyribonucleoside. One of the nitrogen bases joined to deoxyribose (deoxyribose + base).

Dextrose. See Glucose.

Diauxic growth. The phenomenon in which there are two distinct exponential phases separated by a short lag period.

Diploid. A cell which contains two sets of each type of chromosome. (See also Chromosomes).

DNA Polymerase. Enzyme that catalyzes the formation of DNA from deoxyribonucleoside triphosphates utilizing existing DNA as a template.

Dominant. In genetics, referring to a gene which is always expressed phenotypically, that is, it always masks or suppresses the expression of the other allele.

Elasticity (of the air). A poorly defined property of air which its proponents claimed was necessary for spontaneous generation.

Electron microscope. An instrument invented in 1937 that uses beams of electrons to visualize material.

Endoplasmic reticulum. An extensive system of membranes found in the cytoplasm of eucaryotic cells, often coated with ribosomes.

Entropy. A measure of the degree of disorder or randomness of a system. The entropy of a system can only decrease, become more ordered, if energy is supplied.

Enzymes. Proteins which act as catalysts. Catalyses by enzymes are characterized by great efficiency and high specificity. Most enzymes catalyze only one type of chemical reaction.

Episome. A genetic element that can exist either free or connected to a chromosome. An example is a temperate phage in a lysogenic bacterium.

Escherichia coli (E. coli). A nonpathogenic bacterium found in the intestines of man; it is easy to grow and manipulate in the laboratory and thus has been used widely in the study of molecular biology.

Eucaryotic cell. Typical of all cell types except bacteria and blue-green algae, consisting of a well-defined nucleus separated from the cytoplasm by a nuclear membrane and structurally differentiated cytoplasm.

Exponential phase. That part of the growth curve during which the rate of increase is maximum and constant.

F (Fertility) factor. An episome symbolized by F^+ which determines the sex of a bacterium. The presence of F^+ in the cell makes it a male.

Feedback inhibition. The inhibition of the activity of the first enzyme of a biosynthetic pathway by the endproduct of that pathway.

Fermentation. The energy-yielding enzymatic breakdown of nutrients in the absence of oxygen.

Flagella. Cell appendages which are responsible for motility in many microorganisms.

Free energy. That part of the total energy which can perform work at a fixed temperature.

β-Galactosidase. An enzyme which catalyzes the splitting of lactose into glucose and galactose; it is the best studied example of an inducible enzyme.

Galactoside permease. An enzyme located on the cell surface which regulates the entrance of lactose and related sugars into the cell.

Gamete. A haploid germ cell.

Gelatin. A protein obtained from skin, bones, tendons, and so on. It is soluble in boiling water; on cooling, it solidifies to form a transparent gel.

Gene. Part of the hereditary material located in the chromosome. In modern terms a gene can be considered a segment of DNA carrying the information for a single protein.

Genetic code. The four nitrogenous bases (adenine, guanine, cytosine, and thymine in DNA and adenine, guanine, cytosine, and uracil in RNA) constitute an alphabet of four letters; the sequence of these bases in DNA and subsequently RNA forms a code which contains the information for synthesizing specific proteins. This code is now known to consist of three-letter words (Table 6.3), each of which is the code for a specific amino acid. In this way the sequence of amino acids in proteins is determined by the sequence of bases in DNA.

Genotype. The sum of the genes. The genetic makeup of an organism, as contrasted with the characteristics manifested by the organism (phenotype).

Germination. Resumption of growth by spores or other resting cells.

Glucose. A simple sugar having the formula $C_6H_{12}O_6$; also called dextrose and grape sugar.

Glycogen. A large polysaccharide composed of glucose units; it is the major storage product in animal cells.

Glycolysis. One type of fermentation in which glucose is broken down to lactic acid.

Gratuitous inducer. A substance which provokes the synthesis of specific enzyme(s) but which itself is not a substrate.

Growth. The orderly increase of all cellular constituents, leading to an accurate duplication of the existing pattern.

Growth curve. A plot of the number of cells as a function of time; the general shape of the curve is characteristic of all living systems.

Guanine. A purine molecule found in both RNA and DNA (see Figure 1.11).

Haploid. A cell which has only one copy of each type of chromosome.

Heterozygous. In diploid cells, the situation in which the two members of a pair of genes located on homologous chromosomes are different.

Hfr (*High frequency of recombination*). Mutant strains of *E. coli* which show unusually high frequencies of recombination.

Homogentisic acid. An organic compound which is formed in the body from the amino acids phenylalanine and tyrosine. Patients with alkaptonuria cannot degrade this compound and thus excrete it in their urine. (For chemical formula, see Section 6·1.)

Homozygous. In diploid cells, the situation is which the two members of a pair of genes located on homologous chromosomes are identical.

Host cell. A cell which is used for the growth and multiplication of a virus.

Hybrid. An organism resulting from a cross between parents that are genetically unlike in any characteristic. In common usage the term is generally used only when the parents are of different races, varieties, or species.

Independent assortment. Referring to random distribution of genes to the gametes. For example, an organism of genotype *AaBb* will produce equal numbers of the four types of gametes: *AB*, *Ab*, *aB*, and *ab*.

Independent segregation. The first of the Mendelian principles. It states that the two members of each pair of alleles possessed by diploid organisms separate into different gametes.

Inducer. A substance which induces the production of a specific enzyme or group of enzymes by the cell.

Inducible enzymes. Those produced by the cell only in response to specific chemicals in the medium.

Inorganic. Nonliving. In chemistry, molecules which lack carbon.

Inversion. The process of breakage and reunion of the chromosome such that a whole segment is replaced in reverse order.

in vitro. (Latin: in glass). Pertaining to experiments performed with cell-free extracts.

in vivo. (Latin: in life). Pertaining to experiments performed with intact living organisms.

Krebs cycle. A cyclic series of cellular reactions in the breakdown of pyruvic acid to carbon dioxide and the formation of hydrogens for oxidative phosphorylation.

Light reaction. The part of the photosynthetic process that requires light; that is, the first stage in which the radiant energy is captured and transformed into chemical energy.

Linkage. In genetics, the location of two or more genes on the same chromosome so that they are passed on together from parent to offspring.

Lysis. The bursting of a cell by destruction of its cytoplasmic membrane.

Lysogenic bacterium. See Prophage.

Lytic cycle. The sequence of events that lead to the multiplication of viruses and the lysis of the host cell.

Macromolecule. A large molecule, usually built up from small units, for example, proteins, nucleic acids, and polysaccharides.

Meiosis. The process by which diploid somatic cells give rise to haploid gametes. This is brought about by a single duplication of the chromosome followed by two successive reductive divisions.

Messenger RNA (mRNA). RNA manufactured in the nucleus, which moves into the cytoplasm, attaches to ribosomes, and serves as a template for protein synthesis.

Metabolism. All of the closely coordinated, delicately balanced chemical reactions that take place in living cells, entailing the breakdown of various nutrients and the synthesis of cell material.

Micron (μ). A unit of length equal to 10^{-6} meter, 10^{-4} cm, or 10^4 Å.

Minimal medium. A medium containing only those components essential for growth and which the organism cannot synthesize itself.

Mitochondrion. A structure found in the cytoplasm of aerobic animal and plant cells. It is the major site of ATP production.

Mitosis. The process of cell division leading to two genetically identical daughter cells.

Molecule. Any group of atoms forming a close association that remains together long enough to be studied (derived from the Latin words meaning "a small mass"). The atoms comprising the molecule may be alike or different.

Mutagenic agent. Any factor in the environment that increases the probability of a mutation. For example, ultraviolet light is mutagenic since it causes the mutation rate of organisms to rise.

Mutation. Any sudden, heritable change in the structure of the genetic material.

^{15}N. A nonradioactive isotope of nitrogen having an atomic weight of 15; it differs from the usual ^{14}N in having one extra neutron.

Negative feedback inhibition. See Feedback inhibition.

Neurospora crassa. The common bread mold; it is especially useful for biochemical and genetic studies.

Nitrogenous bases. Molecules composed of rings of carbon and nitrogen atoms. Important nitrogen bases in the cell are the purines and pyrimidines. (For chemical formula, see Figure 1.11.)

Nucleic acid. A polymer of nucleotides. (See DNA and RNA.)

Nucleoside. One of the nitrogen bases joined to a sugar (base + sugar).

Nucleotide. One of the nitrogen bases joined to a sugar which is also connected to a phosphate group (base + sugar + phosphate).

Nucleus. That part of the cell which contains the genetic material. In eucaryotic cells the nucleus is bounded by the nuclear membrane.

Operator. A site on the chromosome which regulates the expression of a group of genes; when the repressor combines with the operator region, transcription of the structural genes is inhibited.

Operon. A segment of the chromosome composed of a group of genes whose expression is regulated by a common operator.

Organelle. A general term used in reference to various discrete structures in the cell; for example, mitochondria, ribosomes, and chloroplasts.

Organic. Living. In chemistry, molecules which contain carbon are called organic compounds because they are characteristic of living organisms.

Osmotic work. The energy-requiring process by which cells are able to transport and concentrate molecules into the cell from the environment.

Oxidation. Originally defined as the gain of oxygen or loss of hydrogen. In modern terms, it is a chemical reaction involving the loss of electrons. In some cases the electrons are accepted by oxygen gas.

Oxidative phosphorylation. The process by which oxygen utilization is coupled to ATP production from ADP.

$^{32}P.$ A radioactive isotope of phosphorus with an atomic weight of 32; it is commonly used for labeling nucleic acids in biological experiments.

Panspermia. The theory that life on earth arose from "seeds" (spores) which constantly bombard our planet.

Pasteurization. A heat treatment which kills many of the microbes that are responsible for spoilage or causing diseases without destroying the taste of the food.

Permease. A general name for any enzyme which regulates the entrance of molecules into the cell.

Phages. See Bacteriophages.

Phenotype. The sum of the observable properties of an organism. The phenotype is the result of the interaction of the genotype and the environment.

Photosynthesis. The enzyme catalyzed conversion of light energy into useful chemical energy and use of chemical energy to form sugars and oxygen from carbon dioxide and water.

Photosynthetic phosphorylation. The process by which light energy is utilized to produce ATP from ADP.

Plaque. The clear area on an agar plate which results when viruses cause the lysis of cells in a localized area.

Polymer. A large molecule made up of regular subunits termed monomers; for example, a protein (polymer) is composed of amino acids (monomers).

Polysaccharide. Large carbohydrate molecules composed of many sugars joined together; for example, starch and glycogen.

Procaryotic cell. Typical of bacteria and blue-green algae, characterized by the absence of both nuclear membrane and structurally differentiated cytoplasm.

Prophage. The state of a bacterial virus in which it is integrated into the host chromosome. A bacterium containing a prophage is termed a lysogenic bacterium.

Protein. A polymer of amino acids. Approximately 20 amino acids are commonly found in proteins. The variety of structures and properties of the different amino acids contributes to make proteins the most versatile of biological molecules. Their most important role is that of enzymes.

Prototroph. A microorganism having no nutritional requirements in addition to those of the wild type from which it was derived.

Pure culture. A population of cells that contains a single kind of microorganism and that has originated from a single cell.

Purine. Molecule consisting of two fused rings of five carbon and four nitrogen atoms. Two of the purines, adenine and guanine, serve as bases in RNA and DNA. (For chemical formula, see Figure 1.11.)

Pyrimidine. Molecule consisting of a single ring of four carbon and two nitrogen atoms. Two of the pyrimidines, cytosine and uracil, serve as bases in RNA and two, cytosine and thymine, serve in DNA. (For chemical formula, see Figure 1.11.)

Pyruvic acid. A simple organic compound with the formula $C_3H_4O_3$. Almost all cells are able to break down sugars, such as glucose, to pyruvic acid. The compound can then either be further oxidized to carbon dioxide and water (respiration) or in the absence of oxygen be converted to fermentation products, such as lactic acid and ethanol. In this way pyruvic acid serves as a branch point in energy metabolism.

Radioactive isotope. An isotope with an unstable nucleus so that it emits ionizing radiation.

Recessive. In genetics, referring to the lack of phenotypic expression when a gene is in the presence of its dominant allele.

Recombinants. Offspring which contain a combination of genes not present in either parent, due to crossing over or independent assortment.

Recombination. In genetics, the production of a new genotype in an offspring by independent assortment of genes.

Repression. The inhibition of the formation of enzymes of a biosynthetic pathway by the endproduct of that pathway.

Repressor. With specific regard to the lactose operon, the protein product of the *i* gene which prevents transcription of specific regions of the chromosome.

Respiration. The complete breakdown of nutrient molecules to carbon dioxide and water by aerobic cells.

Ribonuclease (RNase). Enzyme catalyzing degradation of RNA.

Ribonucleic acid (RNA). A linear polymer composed of four nucleotides, adenylic, guanylic, cytidylic, and uridylic acids. In each of these nucleotides the sugar is ribose. All cells contain RNA, whose major function is protein synthesis (see Messenger RNA, Ribosomal RNA, and Transfer RNA). Also, some viruses contain RNA as their genetic material.

Ribosomal RNA (rRNA). The nucleic acid component of ribosomes. It is the major type of RNA in the cell.

Ribosome. Cytoplasmic particle consisting of RNA and protein; it is the site of protein synthesis.

RNA Polymerase. Enzyme that catalyzes the formation of RNA from nucleoside triphosphates utilizing DNA as a template.

[35]*S.* A radioactive isotope of sulfur with an atomic weight of 35; it is commonly used for labeling protein in biological experiments.

Segregation. See Independent segregation.

Somatic cells. The diploid body cells of an organism; those cells other than the germ cells.

Spheroplasts. Bacteria with their cell walls removed. Lacking a rigid cell wall, they become spherically shaped.

Spontaneous generation. The theory that living organisms develop from nonliving matter. Also known as abiogenesis.

Spore. A thick-walled cell capable of surviving adverse environmental conditions.

Starch. A large polysaccharide composed of glucose units; it is the major storage product in plant cells.

Stationary phase. That part of the growth curve in which the population has reached the maximal level that the environment permits; in this phase there is no net increase or decrease in the number of cells.

Sterilization. A process which *completely* destroys or removes all living organisms.

Substrate. Substance acted on by an enzyme. For example, DNA is the substrate for DNase and lactose for β-galactosidase.

Sugar. Simple carbohydrate molecule, generally having a sweet taste. (For some examples, see Figure 1.10.)

Temperate phages. Viruses that infect bacteria; however, instead of destroying the cells, they are maintained within the surviving bacteria in the form of prophages (see Prophage).

Template. The macromolecular mold for the synthesis of another macromolecule. For example, DNA serves as a template (mold) for the synthesis of RNA.

Thymine. A pyrimidine molecule generally found only in DNA (see Figure 1.11).

Transcription. A step in protein synthesis in which part of the DNA is used as a template for the production of a complementary sequence of bases in an RNA chain.

Transduction. Transfer of heritable characters, genes, from one bacterium to another by means of a bacteriophage.

Transfection. A process by which DNA purified from phages enters bacteria which are able to be transformed. Once the viral DNA enters the cell it then reproduces and gives rise to many new complete phages. The process is identical to infection except for the manner in which the DNA penetrates the bacterium.

Transfer RNA (tRNA). One of a group of small RNA molecules which combines with a specific amino acid and ensures that the amino acid lines up correctly on the messenger RNA.

Transformation. The genetic change in a bacterium brought about by absorbing DNA from other strains of bacteria.

Translation. The steps in protein synthesis which take place on the ribosome whereby information in mRNA is utilized to direct the synthesis of a specific protein.

Translocation. In genetics, a type of mutation in which a segment of one chromosome becomes attached to a nonhomologous chromosome.

Trypsin. An enzyme that breaks down protein.

Tyndallization. A sterilization technique, consisting of heating to 100°C for 30 minutes on each of three successive days.

Uracil. A pyrimidine molecule generally found only in RNA (see Figure 1.11).

Vegetative force. A poorly defined substance which the proponents of spontaneous generation claimed was necessary for the formation of living organisms from nonliving matter.

Virulent phages. Viruses which infect bacteria and always lyse the cells during the process of producing progeny phages.

Virus. An infectious agent smaller than a bacterium which lacks cell structure and which can only multiply inside a host cell; viruses contain either DNA or RNA.

Yeast extract. Prepared by extracting dead yeast cells with water and evaporating the liquid to dryness. It is a rich source of vitamins and other growth factors.

Zygote. The diploid cell which is formed by the union of male and female sex cells.

INDEX

DNA Polymerase, 152–156, 221
Dobell, Clifford, 3
Drosophilia melanogaster, 107–116, 160

Einstein, Albert, 34, 71, 199
Electric work, 82
Electron microscope, 38, 221
Elements, 21
Endoplasmic reticulum, 50, 222
Entropy, 72–73, 222
Enzymes, 222
    catalysts, 78–81
    DNA polymerase, 152–156
    DNase, 137–138
    electron micrographs, 69
    genetic control, 159–168
    Krebs cycle, 87
    pepsin, 141
    regulation, 183–201
    RNase, 137–138
    trypsin, 137–138
Episome, 123, 222
Ethyl alcohol, 73–78, 84, 86
Eucaryotic cell, 40, 222
Escherichia coli (E. coli), 222
    DNA base composition, 144
    DNA biosynthesis, 154
    growth, 115–116, 175–185
    Hfr, 119
    mutation, 113, 115–116
    ribosomes, 166
    sex, 116–123
Exponential phase, 177–181

F (Fertility) factor, 119–123
Feedback inhibition, 193–197
Filtration, 12–14
Flagella, 48, 222
Flexner, A., 147
Fermentation, 222
    beer and wine, 74, 85–86
    controversy, 73–78
    metabolism, 81, 83–86
Fox, Sydney, 27

Franklin, Rosalind, 145
Free energy, 72–73, 222

β-Galactosidase, 185–192, 222
Galactoside permease, 187–192, 222
Galactoside transacetylase, 187–192
Garrod, Sir Archibald E., 158–159
Gelatin, 17, 27, 222
Genes, 222
    bacteria, 116–123
    chemical composition, 136–140
    chromosomes, 107–111
    enzyme formation, 159–168
    Mendel's experiments, 100–107
    mutation, 111–116
    regulation, 187–192
    viruses, 130–131
Genetic code, 168–173, 222
Gerhart, John, 197
Germination, 11, 223
Gilbert, Walter, 190–191
Glucose, 24, 26, 176–177, 182–185, 223
    fermentation, 74–79, 84–85
    formation by photosynthesis, 91–96
    formula, 24
Glycine, 23, 26, 27
    formula, 23
Glycogen, 25, 181, 196, 223
Goulian, Mehran, 154
Green, David E., 88
Griffith, Fredrick, 134–136
Growth, 175–185, 223
    of coacervates, 28–29
    diauxic, 182–185
    growth curve, 176–182
    of mutants in defined media, 163–164
    regulation, 185–199
    requirements, 115–116, 175–176
Growth curve, 176–182, 223
Guanine, 25, 26, 142–147, 223
    formula, 25, 142

Haldane, J. B. S., 20
Hales, Stephen, 90
Harden, Arthur, 84